T0210745

The Architectural and Technological Revolution of 5G

José Luiz Frauendorf • Érika Almeida de Souza

The Architectural and Technological Revolution of 5G

 Springer

José Luiz Frauendorf
Ouro Fino, MG, Brazil

Érika Almeida de Souza
São Paulo, SP, Brazil

The translation was done with the help of artificial intelligence (machine translation by the service DeepL.com). A subsequent human revision was done primarily in terms of content.

ISBN 978-3-031-10652-1   ISBN 978-3-031-10650-7   (eBook)
https://doi.org/10.1007/978-3-031-10650-7

This Springer imprint is published by the registered company Springer Nature Switzerland AG
The registered company address is: Gewerbestrasse 11, 6330 Cham, Switzerland

I dedicate this book to a little-known Brazilian hero. He was born in Piracicaba, in the interior of the State of São Paulo, Brazil, in the middle of the nineteenth century, more precisely on October 26, 1853. His name: **Evaristo Conrado Engelberg**, *son of Swiss immigrants. A mechanical engineer by profession, he developed something important, the coffee and rice hulling and separation machine, at a time when Brazil and the State of São Paulo depended economically and fundamentally on coffee culture, considered the green gold. At that time, coffee represented 56% of all Brazilian exports. Engelberg became so famous that Brazilian Emperor Dom Pedro II went personally to Piracicaba to meet him. He became a member of the Paris Academy of Inventors and was awarded with the gold medal. He set up factories in Brazil and in the USA to produce his invention. He didn't become rich, probably because he didn't have enough capital to participate in the investments of the companies that he helped to create and that carried his name. He had to leave the companies, but not before having the satisfaction of seeing his invention patented in the USA on August 4, 1914. He died in 1932, and today he is only remembered in a few books that tell the history of industrialization in Brazil. I am very proud to be his great-grandson. José Luiz Frauendorf*

Foreword

I met Frauendorf when I took over the direction of DirecTV Brazil in 1997. At that time, he was the general director of operations of TVA, a Pay-TV company. Soon after, he became an independent consultant, but continued to work for Editora Abril. Later, as general director of TVA, I had the pleasure of interacting with him for many years and always learned a lot in our conversations and debates. In fact, Frauendorf and Virgílio Amaral, TVA's Technology Director, were very patient, teaching me everything I could learn about wireless technology.

Frauendorf then became general director of Neotec, the Brazilian association of MMDS (Multichannel Multipoint Distribution Service) operators. This association was formed by initiative of the brothers Lins de Albuquerque, Hermano, and Carlos André, from TV Filme Brasília. Neotec was created with the objective of searching in the international market new digital technologies that would allow the expansion of services provided by operators that held the MMDS spectrum concession. The operators, around 2001, were limited, providing pay TV service over the air using the 2.5 GHz band and had great difficulty competing with operators that were providing service using coax cable. At that time everything was analog, but digitalization was very close.

In addition to TVA, TV Filme, Acom, Teleserv, and MMDSC made up a very united team willing to fight for the right to use the spectrum for any other possible service. Carlos André Lins de Albuquerque, representing TV Filme, João Reino and Mário de Paula for Acom, Gisele Gomes for Teleserv, and Odilon Silva for MMDSC completed the team that worked together to optimize the use of the 2.5 GHz spectrum. In leading the association, Frauendorf was tireless in searching new technologies and regulatory alternatives.

Acom had already launched its digital service in Natal and served as a "guinea pig" for the digitization of video services, but this was not enough. MMDS had to provide access to the Internet. TV Filme had launched the provision of access using an MMDS channel for the downstream, and the return, upstream, was done by telephone line. Soon, the services evolved to DOCSIS modems, the same used by cable TV operators. But the team wanted something innovative.

TV Filme, located in the USA some companies that developed specific technologies for wireless. Two of them, Navine and NextNet, were interested in the Neotec project and were willing to perform field tests in Brazil. The Brazilian city of Belo Horizonte, where TV Filme had a concession, was chosen as the ideal stage for the tests. Effectively, after overcoming all the importation and licensing bureaucratic procedures, the equipment arrived in Brazil and the tests could begin in early 2003.

Navine used TD-SCDMA modulation, which had been employed initially in China, like W-CDMA modulation used in 3G. NextNet came with a technological innovation, OFDM modulation, the same used in digital television. OFDM's superiority was verified right at the beginning of the tests that lasted many months. A van was traveling the central streets of BH receiving and transmitting video signals and allowing Internet access, while the connection was always maintained, provided by handover between the three cells installed, covering each an area of 2 to 5 km radius. The success was impressive. The speeds achieved for that time were quite satisfactory: 2 Mbps downstream and 0.8 Mbps upstream.

All operators were invited to witness the success of the tests, as well as, of course, the staff of Anatel, the Brazilian Telecommunications Agency. The then board member of Anatel, Antônio Carlos Valente, soon warned: *Hire a good lawyer to defend your rights.* This was how Dr. Elinor Cotait came to represent Neotec in all the battles for the rights of operators to use the spectrum.

We are very proud of all the work developed by the MMDS operators' association, which was, at that time, recognized internationally.

The results of the BH tests were presented by Frauendorf at a WCA (Wireless Communications Alliance) conference in Palo Alto, California, in January 2004. Among the influential and recognized people in the market who attended the congress were Sean Maloney, VP of Intel, and Barry J. West, CTO of Nextel Communications (Sprint). The invitation for Neotec to be part of the WiMAX Forum, where the proposed technologies solutions were in discussions, was immediate. Intel and Samsung had an agreement to create WiMAX, which would be a system similar to Wi-Fi for outdoor use and intended to be the fourth generation of cellular systems. Shortly after, the WiMAX/4G—Global Development Committee (GDC) was created within the WCA and Neotec was indicated to the co-chair position alongside a representative from Sprint.

All this development motivated Samsung and Motorola to promote WiMAX tests in Brazil, which took place in São Paulo in 2005, and I had the pleasure to follow. Interesting to note that, in the case of Samsung, São Paulo was the second city where field tests were conducted, right after the tests in Seoul, Korea, and before the tests performed by Sprint in the USA, demonstrating that Brazil was really committed to the development of the new technology.

WiMAX didn't take off, but it was the embryo of what we now know as LTE/4G, which in turn was the embryo of 5G, which will enable the transformation of various industries and services.

I was very happy when Frauendorf called me saying that he was launching a book 20 years after Neotec was founded, writing about the technology whose birth and development worldwide he had the opportunity to follow. I was very proud to have participated in an organization that was developing pioneering and innovative work in a country that does not always value initiatives like this.

It is a privilege for me to be able to tell a little of this story and, at the same time, to register how important was the participation of Virgílio, Carlos André, Odilon, João Reino, Mário de Paula, and Gisele in Neotec. The partnership with the operators allowed Frauendorf, as general director of the Association, to have the opportunity to follow the evolution of cellular technology in Brazil and in the world.

To Frauendorf, my special thanks for his dedication and competence in conducting this important project, which, with the arrival of 5G, will bring so many benefits to society.

Chairman of the Board of Directors IBGC (Brazilian Institute of Corporate Governance)

Leila Abraham Loria,
São Paulo, Brazil

Preface

I stayed almost 10 years away from my professional origin, taking care of other interests not related to telecommunications.

I had the privilege of working with WiMAX technology, an innovative technology, evolution of Wi-Fi for outdoor networks and precursor of LTE/4G. It was an incredible learning experience. Unfortunately, the technology didn't take off and I got frustrated.

With the pandemic, I had to retire from the day-to-day work I've done for over 50 years. I became interested in 5G just because I soon realized there was something disruptive on the horizon. The beginning was difficult. The number of acronyms confused me. I felt like I didn't understand anything. In part, this was due to my total lack of knowledge about the architecture of the LTE system, used in both 4G and 4.5G. In fact, I knew very little about the previous systems, as I had not participated in any project that required me to deepen my knowledge in this segment. I was heavily involved in video digitization for a long time, an activity that required me to be very focused.

Gradually, I understood the logic that guided the development of 5G and realized the greatness of this revolutionary technology that will certainly radically change the telecommunications scenario in a very short time. As I understood it, I realized how innovative it is.

As I had done since I was in college, I made summaries of everything I learned. I acted more like a journalist studying a new subject. I hope this book can be of value and inspire those who wish to understand a little about cellular systems and anticipate everything that 5G will mean for users, network operators, and, above all, equipment suppliers, service providers, and application developers.

Basically, the great objective of this book is to divulge new opportunities that will be available to those who are alert.

Three technological quantum leaps are occurring almost simultaneously. The most advanced is photovoltaics, which will greatly change the way electricity is generated and consumed.

The second is exactly 5G, the subject of this book.

The third is electric and hydrogen-powered vehicles, which are already beginning to unbalance a long-standing hegemony. A single indicator is enough to demonstrate their transformative character. While a conventional car is composed of 30,000 parts, electric cars require only 6,000 to be assembled. It is just a matter of time and production scale. Countless opportunities will come around.

But, as more heads think much better than just one, I went searching help from incredible people who shared my professional history, to help me in the task I had set myself. Érika Almeida de Souza collaborated a lot with the preliminary texts, inserting points that required further clarification. She ended up becoming my co-author for her valuable contributions. Erich Baumeier, with an excellent professional background in IT, information technology, also collaborated a lot with comments and suggestions, mainly in the segment he dominates. Working with more people makes the task even more interesting and much more enjoyable.

Our intention is to write something that will help all readers interested in 5G to be more knowledgeable and aware of the new opportunities. I hope we can achieve our goals. Happy reading!

Important note: for those less familiar with some basic concepts used in telecommunications and computer networks, we advise reading three additional chapters: "Digital Modulation," "Computer Networks," and "AI—Artificial Intelligence/ML—Machine Learning."

Ouro Fino, MG, Brazil   José Luiz Frauendorf
April 2022

Contents

About the Authors

José Luiz Frauendorf was born in São Paulo, Brazil. He got his degree in electronic engineer from the Institute Mauá of Technology. He started his professional career at AEG Telefunken, in Backnang, Germany, where he worked on the development of telecommunication systems, including coaxial cable systems and microwaves. Later, he was transferred to Brazil with the responsibility of implementing at the local subsidiary the manufacturing of public telephony systems which included voice channels multiplexing systems, microwave systems, and monitoring and supervision systems for telecommunications stations. With the acquisition by Grupo Docas de Santos of ELEBRA Eletrônica Brasileira, he was invited to take over the product development area of ELEBRA Informática. Subsequently, he held the position of industrial and technical director of ELEBRA Computers until its acquisition by Digital Equipment Corporation, when he became the plant manager of the Brazilian subsidiary. He participated in TVA—Grupo ABRIL Television System, having been its general director of operations. After that, he became general director of NEOTEC, an association created to develop technologies to provide multimedia services using the 2.5GHz spectrum. Frauendorf coordinated tests of systems providing broadband services using WCDMA and OFDM modulation systems, and later WiMAX technology, in several Brazilian cities. In 2010, he participated in the turnaround of a family company, from which he withdrew when it was sold in 2014. After his experience in the polymer segment, he developed a new technology for electrically insulating gloves using synthetic materials. His invention was patented in several countries. Since 2020, he has been dedicating himself in writing books and disseminating news in the areas of new technologies.

Érica Almeida de Souza got her degree in telecommunications engineering and postgraduate from INATEL (National Institute of Telecommunications). She participated in several training courses in telephony and data transmission at Huawei, both in China and in Brazil. She started her professional career at the former CFLCL, now Energisa. She also worked at Panasonic do Brasil and later at Huawei, where she had the opportunity to promote training courses in Brazil and Latin America. She is currently caring out advance training courses at IWF Training and Consulting.

Abbreviations

3GPP	Third Generation Partnership Project
AAA	Authentication, Authorization and Accounting
ACIR	Adjacent Channel Interference Ratio
ACK	Acknowledgement
ACLR	Adjacent Channel Leakage Ratio
ACS	Adjacent Channel Selectivity
ADC	Analog-to Digital Conversion
ADSL	Asymmetric Digital Subscriber Line
AKA	Authentication and Key Agreement
AM	Acknowledged Mode
AMBR	Aggregate Maximum Bit Rate
AMD	Acknowledged Mode Data
AMR	Adaptive Multi-Rate
AMR-NB	Adaptive Multi-Rate Narrowband
AMR-WB	Adaptive Multi-Rate Wideband
ARP	Allocation Retention Priority
ASN	Abstract Syntax Notation
ATB	Adaptive Transmission Bandwidth
AWGN	Additive White Gaussian Noise
BB	Baseband
BCCH	Broadcast Control Channel
BCH	Broadcast Channel
BE	Best Effort
BEM	Block Edge Mask
BICC	Bearer Independent Call Control Protocol
BLER	Block Error Rate
BO	Backoffice
BOM	Bill of Material
BPF	Band Pass Filter
BPSK	Binary Phase Shift Keying
BS	Base Station

BSC	Base Station Controller
BSR	Buffer Status Report
BT	Bluetooth
BTS	Base Transceiver Station
BW	Bandwidth
CBR	Constant Bit Rate
CCCH	Common Control Channel
CCE	Control Channel Element
CDD	Cyclic Delay Diversity
CDF	Cumulative Density Function
CDM	Code Division Multiplexing
CDMA	Code Division Multiple Access
CIR	Carrier to Interference Ratio
CLM	Closed Loop Mode/Close Loop Methodology
CM	Cubic Metric
CMOS	Complementary Metal Oxide Semiconductor
CoMP	Coordinated Multiple Point
CP	Cyclic Prefix
CPE	Common Phase Error
CPICH	Common Pilot Channel
CQI	Channel Quality Information
CRC	Cyclic Redundancy Check
C-RNTI	Cell Radio Network Temporary Identifier
CS	Circuit Switched
CSCF	Call Session Control Function
CSFB	Circuit Switched Fallback
CSI	Channel State Information
CT	Core and Terminals
CTL	Control
CW	Continuous Wave
DAC	Digital to Analog Conversion
DARP	Downlink Advanced Receiver Performance
D-BCH	Dynamic Broadcast Channel
DC	Direct Current
DCCH	Dedicated Control Channel
DCH	Dedicated Channel
DC-HSDPA	Dual Cell (Dual Carrier) HSDPA
DCI	Downlink Control Information
DCR	Direct Conversion Receiver
DCXO	Digitally Compensated Crystal Oscillator
DD	Duplex Distance
DFCA	Dynamic Frequency and Channel Allocation
DFT	Discrete Fourier Transform
DG	Duplex Gap
DL	Downlink

DL-SCH	Downlink Shared Channel
DPCCH	Dedicated Physical Control Channel
DR	Dynamic Range
DRX	Discontinuous Reception
DSP	Digital Signal Processing
DTCH	Dedicated Traffic Channel
DTM	Dual Transfer Mode
DTX	Discontinuous Transmission
DVB-H	Digital Video Broadcast–Handheld
DwPTS	Downlink Pilot Time Slot
E-DCH	Enhanced DCH
EDGE	Enhanced Data Rates for GSM Evolution
EFL	Effective Frequency Load
EFR	Enhanced Full Rate
EGPRS	Enhanced GPRS
E-HRDP	Evolved HRPD (High-Rate Packet Data) network
EIRP	Equivalent Isotropic Radiated Power
EMI	Electromagnetic Interference
EPC	Evolved Packet Core
EPDG	Evolved Packet Data Gateway
ETU	Extended Typical Urban
E-UTRA	Evolved Universal Terrestrial Radio Access
EVA	Extended Vehicular A
EVDO	Evolution Data Only
EVM	Error Vector Magnitude
EVS	Error Vector Spectrum
FACH	Forward Access Channel
FCC	Federal Communications Commission
FD	Frequency Domain
FDD	Frequency Division Duplex
FDE	Frequency Domain Equalizer
FDM	Frequency Division Multiplexing
FDPS	Frequency Domain Packet Scheduling
FE	Front End
FFT	Fast Fourier Transform
FM	Frequency Modulated
FNS	Frequency Non-Selective
FR	Full Rate
FRC	Fixed Reference Channel
FS	Frequency Selective
GB	Gigabyte
GBR	Guaranteed Bit Rate
GDD	Group Delay Dispersion
GERAN	GSM/EDGE Radio Access Network
GGSN	Gateway GPRS Support Node

GMSK	Gaussian Minimum Shift Keying
GP	Guard Period
GPON	Gigabit Passive Optical Network
GPRS	General Packet Radio Service
GPS	Global Positioning System
GRE	Generic Routing Encapsulation
GSM	Global System for Mobile Communications
GTP	GPRS Tunneling Protocol
GTP-C	GPRS Tunneling Protocol, Control Plane
GUTI	Globally Unique Temporary Identity
GW	Gateway
HARQ	Hybrid Automatic Repeat and Request
HB	High Band
HD-FDD	Half Duplex Frequency Division Duplex
HFN	Hyper Frame Number
HII	High Interference Indicator
HO	Handover
HPBW	Half Power Beam Width
HPF	High Pass Filter
HPSK	Hybrid Phase Shift Keying
HRPD	High-Rate Packet Data
HSDPA	High Speed Downlink Packet Access
HS-DSCH	High Speed Downlink Shared Channel
HSGW	HRPD Serving Gateway
HSPA	High Speed Packet Access
HS-PDSCH	High Speed Physical Downlink Shared Channel
HSS	Home Subscriber Server
HS-SCCH	High Speed Shared Control Channel
HSUPA	High Speed Uplink Packet Access
IC	Interference Cancellation
ICI	Inter-carrier Interference
ICIC	Inter-Cell Interference Coordination
ICS	IMS Centralized Service
ID	Identity
IETF	Internet Engineering Task Force
IFFT	Inverse Fast Fourier Transform
IL	Insertion Loss
IMD	Intermodulation Distortion
IMS	IP Multimedia Subsystem
IMT	International Mobile Telecommunications
IoT	Internet of Things
IOT	Inter-Operability Testing
IP	Internet Protocol
IR	Image Rejection
IRC	Interference Rejection Combining

ISD	Inter-site Distance
ISDN	Integrated Services Digital Network
ISI	Inter-Symbol Interference
IWF	Interworking Function
LAI	Location Area Identity
LB	Low Band
LCID	Logical Channel Identification
LCS	Location Services
LMA	Local Mobility Anchor
LMMSE	Linear Minimum Mean Square Error
LNA	Low Noise Amplifier
LO	Local Oscillator
LOS	Line of Sight
LTE	Long-Term Evolution
MAC	Medium Access Control
MAP	Mobile Application Part
MBMS	Multimedia Broadcast Multicast System
MBR	Maximum Bit Rate
MCH	Multicast Channel
MCL	Minimum Coupling Loss
MCS	Modulation and Coding Scheme
MGW	Media Gateway
MIB	Master Information Block
MIMO	Multiple Input Multiple Output
MIP	Mobile IP
MIPI	Mobile Industry Processor Interface
MIPS	Million Instructions Per Second
MM	Mobility Management
MME	Mobility Management Entity
MMSE	Minimum Mean Square Error
MPR	Maximum Power Reduction
MRC	Maximal Ratio Combining
MSC	Mobile Switching Center
MSC-S	Mobile Switching Center Server
MSD	Maximum Sensitivity Degradation
MU	Multiuser
NACC	Network Assisted Cell Change
NACK	Negative Acknowledgement
NAS	Non-access Stratum
NAT	Network Address Translation
NB	Narrowband
NF	Noise Figure
NMO	Network Mode of Operation
NRT	Non-real Time
OFDM	Orthogonal Frequency Division Multiplexing

OFDMA	Orthogonal Frequency Division Multiple Access
OI	Overload Indicator
OLLA	Outer Loop Link Adaptation
OOB	Out of Band
OOBN	Out-of-Band Noise
O&M	Operation and Maintenance
PA	Power Amplifier
PAPR	Peak to Average Power Ratio
PAR	Peak-to-Average Ratio
PBR	Prioritized Bit Rate
PC	Personal Computer
PC	Power Control
PCC	Policy and Charging Control
PCCC	Parallel Concatenated Convolutional Coding
PCCPCH	Primary Common Control Physical Channel
PCFICH	Physical Control Format Indicator Channel
PCH	Paging Channel
PCI	Physical Cell Identity
PCM	Pulse Code Modulation
PCRF	Policy and Charging Rules Function
PCS	Personal Communication Services
PDCCH	Physical Downlink Control Channel
PDCP	Packet Data Convergence Protocol
PDF	Probability Density Function
PDI	Precoding Matrix Indicator
PDN	Packet Data Network
PDSCH	Physical Downlink Shared Channel
PDU	Protocol Data Unit
P-GW	Packet Data Network Gateway
PHICH	Physical HARQ Indicator Channel
PHY	Physical Layer
PLL	Phase Locked Loop
PLMN	Public Land Mobile Network
PMI	Precoding Matrix Index
PMIP	Proxy Mobile IP
PN	Phase Noise
PRACH	Physical Random-Access Channel
PRB	Physical Resource Block
PS	Packet Switched
PSD	Power Spectral Density
PSS	Primary Synchronization Signal
PUCCH	Physical Uplink Control Channel
PUSCH	Physical Uplink Shared Channel
QAM	Quadrature Amplitude Modulation
QCI	QoS Class Identifier

QN	Quantization Noise
QoS	Quality of Service
QPSK	Quadrature Phase Shift Keying
RACH	Random Access Channel
RAD	Required Activity Detection
RAN	Radio Access Network
RAR	Random Access Response
RAT	Radio Access Technology
RB	Resource Block
RBG	Resource Block Group
RF	Radio Frequency
RI	Rank Indicator
RLC	Radio Link Control
RNC	Radio Network Controller
RNTP	Relative Narrowband Transmit Power
ROHC	Robust Header Compression
RRC	Radio Resource Control
RRM	Radio Resource Management
RS	Reference Signal
RSCP	Received Symbol Code Power
RSRP	Reference Symbol Received Power
RSRQ	Reference Symbol Received Quality
RSSI	Received Signal Strength Indicator
RT	Real Time
RTT	Round Trip Time
RV	Redundancy Version
SA	Stand Alone
SAE	System Architecture Evolution
SAIC	Single Antenna Interference Cancellation
S-CCPCH	Secondary Common Control Physical Channel
SC-FDMA	Single Carrier Frequency Division Multiple Access
SCH	Synchronization Channel
SCM	Spatial Channel Model
SCTP	Stream Control Transmission Protocol
SDU	Service Data Unit
SE	Spectral Efficiency
SEM	Spectrum Emission Mask
SF	Spreading Factor
SFBC	Space Frequency Block Coding
SFN	System Frame Number
SGSN	Serving GPRS Support Node
S-GW	Serving Gateway
SIB	System Information Block
SID	Silence Indicator Description
SIM	Subscriber Identity Module

SIMO	Single Input Multiple Output
SINR	Signal to Interference and Noise Ratio
SMS	Short Message Service
SNDR	Signal to Noise and Distortion Ratio
SNR	Signal-to-Noise Ratio
SON	Self-Optimized Networks
SON	Self-Organizing Networks
SR	Scheduling Request
S-RACH	Short Random-Access Channel
SRB	Signaling Radio Bearer
S-RNC	Serving RNC
SRS	Sounding Reference Signals
SR-VCC	Single Radio Voice Call Continuity
SSS	Secondary Synchronization Signal
S-TMSI	S-Temporary Mobile Subscriber Identity
SU-MIMO	Single User Multiple Input Multiple Output
S1AP	S1 Application Protocol
TA	Tracking Area
TBS	Transport Block Size
TD	Time Domain
TDD	Time Division Duplex
TD-LTE	Time Division Long-Term Evolution
TD-SCDMA	Time Division Synchronous Code Division Multiple Access
TM	Transparent Mode
TPC	Transmit Power Control
TRX	Transceiver
TSG	Technical Specification Group
TTI	Transmission Time Interval
TU	Typical Urban
UDP	User Datagram Protocol
UE	User Equipment
UHF	Ultra-High Frequency
UICC	Universal Integrated Circuit Card
UL	Uplink
UL-SCH	Uplink Shared Channel
UM	Unacknowledged Mode
UMD	Unacknowledged Mode Data
UMTS	Universal Mobile Telecommunications System
UpPTS	Uplink Pilot Time Slot
USB	Universal Serial Bus
USIM	Universal Subscriber Identity Module
USSD	Unstructured Supplementary Service Data
UTRA	Universal Terrestrial Radio Access
UTRAN	Universal Terrestrial Radio Access Network
VCC	Voice Call Continuity

VCO	Voltage Controlled Oscillator
VDSL	Very High-Bit Rate Digital Subscriber Line
VLR	Visitor Location Register
V-MIMO	Virtual MIMO
VoIP	Voice over IP
WCDMA	Wideband Code Division Multiple Access
WG	Working Group
WLAN	Wireless Local Area Network
WRC	World Radiocommunication Conference
X1AP	X1 Application Protocol

Chapter 1
Remembering the Technological Revolutions

Humankind takes cultural and technological leaps from time to time. The first one was during the Renaissance, more specifically between the fifteenth and sixteenth centuries. But it was only with the invention of the steam engine, already in the eighteenth century, that animal traction, the power of windmills and water wheels, and even with the progressive abolition of slavery, that humanity had a new option of driving force, met the *First Industrial Revolution*, and entered the Age of Steam and machinery. It was then that the first railroads were born. This revolution brought some important consequences, such as the coming of the people from the countryside to the cities, and the replacement of craft work by industrial work, especially in the textile industry. Unfortunately, the emergence of capitalism, creating the bourgeois and proletarian classes, would generate social conflicts in a very short time. The main source of energy was coal, which caused the increase of atmospheric pollution.

The *Second Industrial Revolution* did not take long to emerge, already at the beginning of the twentieth century. It was characterized by the introduction of electricity, telegraph, and telephone, allowing greater communication between peoples. It was indeed a revolution, especially with the possibility of displacement of people and the intensification of the use of railroads. It became a landmark of the internalization of the American territory, beginning a new era of development for its population. The steel industry allowed the construction of ships and the consequent impulse to maritime navigation. The radio, the cinema, the airplane, the combustion engine, and the plastic were also born, and the concept of mass production with Ford was established. The process was totally verticalized and standardized, there was no room for customization. The oligopolies and transnational companies emerged, with a strong concentration of economic power. The chemical and mechanical industries gained prominence. The airplane began to emerge as a means of transportation, competing with railroad and naval transportation. Oil and electricity replaced partially coal as sources of energy generation.

© The Author(s), under exclusive license to Springer Nature Switzerland AG 2023
J. L. Frauendorf, É. Almeida de Souza, *The Architectural and Technological Revolution of 5G*, https://doi.org/10.1007/978-3-031-10650-7_1

The *Third Industrial Revolution* followed soon after the end of World War II, during which several researches and discoveries took place. It was the beginning of mass production and the popularization of consumption. Japan revolutionized the industrial production model with the *just in time* system. Large computers appeared and were soon to be replaced by PCs. The internet facilitated communications, and systems began to change from being analog to digital. Robots were born. The labor force demanded qualification and specialization, and the structure of companies became complex. Technological poles were created. Universities and research centers generate knowledge that was after made available to the industry. Outsourcing and procurement emerged on a global level, as did globalization and flexible production processes. Standardization allowed gain of scale, reducing production costs and massification. The demand for a communication system to support these needs increased. Agribusiness was created to meet the demand for food, and the need generated by consumption was supplied by transgenic seeds. The focus ended up being on innovation, with nuclear energy gaining space as an energy player to ensure demand.

During this period, we had the so-called *Information Age*, represented by the junction of telecommunication and the internet. This technological era began in 1950 with the invention of the computer, evolved, and in 1960 enabled the creation of the ARPANET, forerunner of the internet. In 1970, the standardization of the data information exchange protocol (TCP/IP) allowed the networks to take off. Finally, in 1980, with the introduction of http (Hypertext Transfer Protocol) and protocols that allow the exchange of information such as HTML (HyperText Markup Language) and the www (World Wide Web), which enables access to the addresses of the sites, the foundations for the transformation of the internet into a phenomenon in terms of people communication were structured. We use all these tools without even realizing how much they mean in our lives. This technological revolution has aggravated, on the one hand, the problems of manpower qualification, but it has also opened to the whole society immense sources of knowledge thanks to information search sites. The computer age has begun, in which the main working tools for the most diverse areas of knowledge are algorithms. Clean energy begins to emerge as an alternative to the types of energy used until then. Brazil made an important contribution with the development of bi-fuel or even "tri-fuel" cars (gasoline, ethanol, and gas), but mainly using alcohol. Solar and wind energy are gradually gaining scale and intend to definitively replace polluting sources.

The *Fourth Industrial Revolution*, also known as *Industry 4.0*, emerged from the idea of promoting the computerization of manufacturing and data integration, a concept proposed during the Industrial Fair in Hannover, Germany, in 2011. The concept of intelligent factories was born there, in which machines and equipment could make decisions based on data. Industry 4.0 does not only cover manufacturing units, but rather the entire industrial ecosystem, and is favorable to new business opportunities. The increasing automation predicts that a large part of the workforce will be replaced by robots or cobots (collaborative robots) and makes unrestricted use of new technologies and applications such as Artificial Intelligence (AI), Machine Learning (ML), Data Science, nanotechnology, Internet of Things (IoT), Connected

Cars, Surgical Robots, 3D printers, and many other creative ideas that are appearing almost daily.

The processing capacity of CPUs still limits current technology. While quantum computing does not arrive, CPUs with multiple cores (CPU brains), the cheapening of storage in memory devices with better performance (SSD – Solid State Drive), storage and processing in the cloud (Cloud Storage & Computing), the availability of capacity in local, metropolitan, and long-distance networks all contribute to create this scenario of technological growth. It is the time of Everything Connected, including the most diverse applications, such as industrial and home automation, and wearables. We are living yet another incredible era!

But what are the big technological components that contributed to make it all come true? Let's tell that fascinating story.

It began timidly in 1980, with the first generation 1G of the so-called cellular telephony, which, as the name says, only served to allow telephone conversations still in analog format (AMPS).

In 1990, mobile telephony went digital and thus became the embryo of data transmission within mobile telephony. It is the GSM/CDMA technology, which in addition to telephone calls allowed data transmission at a speed of 64 kbps, like the service provided by fixed telephone service companies that already provided data services – Dial-Up – with the characteristic sound that older people knew very well.

The "hunger" for faster data transmission came via ADSL modems, which fixed-line operators made available for residential use. As cellular service companies could not be left behind, they launched 2.5G networks, which allowed reaching 144 kbps using a new technology, GPRS/EDGE.

In 2000, a new service appeared, the 3G, which allowed, in addition to voice services, data services with up to 2 Mbps. It was already possible to transmit e-mails and download MP3 music. Meanwhile, fixed service providers offered Cable Modem in cable TV networks. Remember that these speeds were all theoretical and in practice generated extensive levels of complaints. There was some progress, and cellular operators launched 3.5G using HSPA+ technology, which even today allows speeds of 10 Mbps. This enabled the introduction of smartphones. It is also important to note that, since 3G, the ITU (International Telecommunication Union) and 3GPP (3rd Generation Partnership Project) have taken an increasingly important role in the standardization of telecommunication systems, working alongside the industry in the design of new networks.

At this point in time, the partnership of Intel and Samsung wanted to innovate and launched a revolutionary technology, WiMAX/WiBRO. INTEL's success with Wi-Fi made it venture into higher flights. The dream was short-lived. The traditional telecommunication companies, manufacturers, and service providers felt threatened and did not accept to leave for a totally new system, totally IP. But they didn't stand still either, soon they got together to launch LTE – Long-Term Evolution. They took advantage of the technological embryo of WiMAX/WiBRO, which used a revolutionary modulation system, OFDM (Orthogonal Frequency Division Multiplex), the same transmission system that had allowed the worldwide launch of fixed and mobile digital TV. This learning process was long and costly.

Organizations, accustomed to voice transmission systems, had difficulty embracing IP networks. The system was launched only to serve data services in 2010 and was called 4G. Meanwhile, they were effectively finishing the development of a new system that would be totally IP and would allow phone calls using VoIP (Voice over IP). This service was named VoLTE (Voice over LTE), consisting of LTE access and the packet-based voice switching system called IMS (IP Multimedia Subsystem), now used all over the world. When it was launched, 4G allowed speeds up to 100 Mbps (3GPP – Release 8), while 4.5G reaches 300 Mbps (3GPP – Release 13). It was only then that the "telephone operators" abandoned once and for all the switching system used in telephone lines and understood that the "telephone service" was no longer the main service they offered, even though it had provided many profits over decades. The world changed and they had to change too!

High profits made it possible for telephone service providers, and especially mobile service providers, to invest in the construction of their networks. With each new technology launched, they were forced to maintain compatibility with old systems, what made their networks more expensive. WiMAX/WiBRO intended to clean up the network architecture without having to worry about legacy systems. It didn't go ahead, but it served to push and take out of the comfort zone the traditionalists, addicted to voice services. The lack of consistent specifications, that allowed the consolidation of services without having to introduce constantly changes on the networks, cost a lot of work and money.

The frequency spectrum used by various technologies had to be shared among the different operators. Around 2010, we joked in Brazil that the relevant authorities were "chopping up the spectrum" in order to create competition. This forced the development of a special technology that allowed to join several parts of different frequency bands, to obtain more bandwidth and thus increase the speed of data transmission, the so-called "Carrier Aggregation". We claimed that "broadband" is the synonym of "wide/ample spectrum", something already demonstrated by the North American Mathematician Claude Shannon, grandfather and precursor of digital systems, back in 1948.

With the arrival of 2020, 5G was emerging. This system is a quantum leap in telecommunication technology and proposes to transmit data, images, videos, music, IoT, in short, any application within the IP standard, enabling a series of new services that are still being created.

It would be wrong to think of 5G solely as a mobile wireless network. It is much more than that, for the simple reason that it is an integral part of the infrastructure that will support the launch of new technologies such as autonomous vehicles, drones, remote sensing devices (IoTs), control of cyber systems (robots, mechanical arms, cobots, etc.), virtual reality applications, augmented reality, games, remote surgery, and other novelties. It should be noted that this type of service demands much more from the network in terms of response speed, capacity (data transmission/reception rate), and number of simultaneous connections than previous versions.

This difference needs to be better explained as the reason does not seem so clear at first glance. You might say, for example, that there have already been tests with autonomous vehicles using 4G technology and that these tests were successful. This is true; however, 5G is not intended to control one vehicle but to allow traffic control of the entire fleet and the traffic control of a given location. The difference in capacity between the two systems is therefore very large.

In some locations where towns are very concentrated, like Brazil, fixed services evolved thanks to the cheapness of fiber optic networks. But this is not a reality in every country where towns are spread over a large geographic area. With fiber, it was possible to reach homes and ensure excellent data services and OTT – Over The Top services, such as Netflix, Prime Video, YouTube, Spotify, etc. This sealed the slow death of conventional pay-TV and paved the way for the massive use of Wi-Fi in home networks. For locations not served by optical fiber, one of the services provided by 5G is FWA (Fixed Wireless Access), which is being called "fiber over the air." The nickname is because the 5G service has fiber quality and allows the user to access *full* HD videos (4 K and 8 K), TV, and games, among all applications that require high transmission quality.

An important point to note is that 5G, at least at the beginning, will be deployed at much higher frequencies than those used in current cellular systems. Higher frequency systems are more susceptible to propagation issues. The attenuation of signals passing through walls, mirrored glass, or any physical obstacle, tends to weaken or even block the transmitted signal. Optionally, 5G offers solutions that use low-power, low-range antennas, especially for small environments – the so-called Femtocells (indoor use only) and Picocells and Microcells (both for indoor and outdoor use).

Meanwhile, 4G is a technology that is widely used and has a large installed base, which guarantees a useful life of a few more years until the installed 4G systems are replaced by 5G. As a result, over the next few years, we will see a combination of 4G, Wi-Fi, and 5G networks. In this case, we say that 5G is a holistic system, since the 5G core is being designed to work with this plurality of systems.

Note, however, that the use of optical fiber will not be eliminated, but increased, since the 5G antennas are connected to the network core mainly through fibers.

5G was recently launched with Dynamic Spectrum Sharing (DSS) technology, which enables dynamic spectrum sharing between 4G and 5G. In other words, DSS allows 5G to operate within the frequency range already made available to 4G. This technology is relatively inexpensive and quick to install. However, the frequency sharing process generates a considerable overhead, not allowing the performance of the DSS system to be the same as expected for the 5G system.

For 5G to be used to its full potential, it was necessary to define a specific operating range for 5G within the frequency spectrum, as the frequency spectrum is already quite busy with other applications (military, medical, mobile telephony, amateur radio, etc.). This process generated a lot of discussion and had to be very well managed by legislators, who had to consider not only what is happening now but, above all, what is coming next.

It may be important, for some readers, to differentiate what is the operating frequency, band, or spectrum band used. The transmission frequency is the frequency at which the system operates, while the band refers to the fraction of the spectrum made available for the operation of a given service.

The most cited frequency for 5G startup is 3.5 GHz, which is within the range from 3 to 6 GHz, or C Band, informally referred to as the low band of 5G. Up to now, the use of this band was reserved for satellite systems and conventional broadcast TV stations, which use the satellite system to distribute their programming. The successful implementation of 5G requires the "cleaning" of this band to avoid interference between systems, which requires strong action by lawmakers. Note that other bands were also considered for the introduction of 5G, looking especially for unused bands within the frequency spectrum and even unlicensed spectrum frequencies. The choice of the 3.5 GHz band has the advantage of already being defined for 5G in many countries, which facilitates the standardization and manufacturing of equipment.

5G services consume a lot of bandwidth. It has been already considered, requiring new spectrum bands for its use. In other words, 5G is an insatiable "glutton" because it takes advantage of all the available spectrum. It is easy to understand such gluttony: it is a system that works with very wide bandwidths. The more bandwidth, the better, because it increases efficiency.

A historical fact occurred in the USA. The auction held by the FCC – Federal Communication Commission in late 2020 raised just over US$80 billion. It exceeded analysts' estimates by almost 100%. This is indicative of the business potential that operators see in 5G technology.

Due to the propagation characteristics of the RF signal, the higher the frequency, the lower the coverage capacity. For this reason, when using higher frequencies, the coverage area of the cells will automatically be reduced and 5G will basically work with small cells, which should have only tens or hundreds of meters of radius. As a result, 5G will probably cause network densification, that is, it will be necessary to increase the number of cells to cover the same area. This will cause the number of installed equipment to skyrocket. It is estimated that the number of 5G cells, depending on the operating frequency, may be multiplied by three or even four times the number used in 4G. This will impact the architecture of cellular networks. This is exactly why the use of DSS is becoming an interesting option for mobile operators, since 4G operates in lower frequencies and allows covering practically the same area without having to add new cells to the system. This will give operators a breathing space, allowing them to save resources in the first 5G installations if latency requirement is not critical.

An interesting phenomenon is taking place. Site sharing companies have already seen the potential market of 5G and are demonstrating, through the launch of shared neutral networks, that thanks to the M&A process, it is possible to join forces with other local/regional providers to exploit this scenario.

In the next chapter, we'll focus on what 5G is promising us. Don't think that everything will come at once. As with previous generations, progress will be gradual and will take time, but the horizon is very promising.

And in the following chapters, we will dedicate ourselves to tell how was the evolution to get where we are, we will go a little deeper into the knowledge about the technologies involved, focusing on the evolution of the various generations of network architecture, language, and protocols used by the equipment that make up the whole system, the modulation system used, and the MIMO antennas with *Beamforming* technology.

We will try to use a minimum of very technical concepts to facilitate reading and understanding without generating discouragement in our readers. In fact, it is not our intention to go into too much detail, but we also do not want to lose the technological vision. We intend this to be an informative text for those who have training in the area, but also sufficiently accessible to those interested in more details about this revolutionary technology.

Chapter 2
Strategic Vision of the New Technology

To understand the disruptive nature of 5G, it is necessary to note that, first, it will not be just a new mobile system, much less a mobile phone system dedicated to voice calls, something that ceased to be a long time ago. Rather, it will be an integrated telecommunications system. 5G New Radio has been designed to support the technological revolution proposed by Industry 4.0 and enable the completely innovative services being created now and even others that may be created in the future.

The way it is being conceived, 5G will require a fiber-optic network structure that will be its backbone. It is this transport network, which interconnects the access network to the Core, that will guarantee its operation, that will make it stand up. It is worth noting that in some places the deployment of fiber optic networks has advanced over the years. Today, cable or fiber networks reach many homes, where the signal is distributed via Wi-Fi. In other words, there is already a high-speed access network inside homes. 5G may improve the existing access and compete with cable operators.

Cellular networks already allow, since 4G, the "handover" to Wi-Fi networks. Whenever we enter a place served by Wi-Fi, and if your device is authorized to access the local Wi-Fi service, the application will switch automatically. The so-called "mobile cellular network" has ceased to be just a mobile network to become an integrated network, capable of providing all services that can be used on any platform, whether smartphone, PC, laptop, or tablet, which makes a lot of sense from the point of view of the user, for whom network issues should be transparent. The idea of total telecom integration would make everyone's life easier.

In the beginning, voice transmission was carried out over a special wired network known as a circuit network, which used technology, protocols, and systems specially developed for landline telephony, which was later adapted for mobile telephony. Over time, this technology was replaced by more competitive techniques that could cope with the flexibility and advances of internet protocols. The solution for voice transport was translated into the Voice over IP system. In VoIP, the signal (voice) is digitized, and voice packets are encapsulated and treated by the network

J. L. Frauendorf, É. Almeida de Souza, *The Architectural and Technological Revolution of 5G*, https://doi.org/10.1007/978-3-031-10650-7_2

in a similar way to what is done with the data packets. Special criteria were adopted to guarantee the quality of voice transmissions, since it is a real-time event. However, it is important to note that within the packet network, "voice" becomes just one more service offered by the internet, competing with all other types of applications, the Apps.

Today, telecom operators provide voice telephony or VoIP services according to the resources available in each region. There are places where we have 2G or 3G networks, which work with traditional voice telephony (circuit switching), and others where there are 4G, or even 5G, networks that use VoIP services. As consumers, we use both services almost interchangeably. Terminals such as smartphones and tablets allow us to easily make phone calls and access internet services (Apps), providing a whole universe of new features. What are we getting at? Soon, there will no longer exist a telephone network. We are entering the era of pure IP networks, and it is under this protocol that practically all applications run.

We might see soon a merger between the mobile cellular network and the internet service as a fully integrated network to be shared by several operators, who will most likely lose their vertical business model characteristics, as still occurs today. In this sense, we already see some progress with the sale of sites where cell towers are installed by specialized companies, as is the case of American Tower. Some operators already use some kind of synergy, such as sharing sites (colocation), infrastructure, access network (RAN), or even idle network resources, which is a good indication. The colocalization of cellular network infrastructures is something that should have been thought of a long time ago. Much investment could have been saved. But we learn by making mistakes, and technological advances will be responsible for breaking monopolies, as we shall see below.

The evolution from the launch of 1G to 4.5G was quite painful, largely due to the need to keep in operation all previous systems, the so-called "legacy systems." The 1G no longer exists, nor could it survive, but the 2G does, despite the years. With technological evolution, traditional operators will invariably have to dismantle old networks, as they are inefficient and costly to maintain. Moreover, the spectrum occupied by these systems is an asset. Spectrum is finite and cannot be manufactured.

The starting point of cellular services was focused on voice service and only became a digital, fully IP service, about 11 years ago with the introduction of 4G VoLTE. The bias of evolution was always the phone service, and it took a long time for pure IP to be understood. But it was important because it allowed to surpass 3G in terms of throughput, bandwidth, and latency. With VoLTE that the system started to operate entirely within the IP environment. 5G is the consolidation of everything learned over the past 40 years (1980–2020) and should become an extraordinary system. But let's not be too optimistic; 5G will only reach full maturity around 2025. Let us focus on the wonders of this new system.

5G can be defined by three main pillars: increased transmission rate, very low latency, and the ability to connect massive amounts of IoT application terminals and mobile phones.

The first characteristic is the *Data Volume* flowing to the user or the data transfer rate, measured in bits per second. In 5G jargon, this is called *eMBB* – Enhanced Mobile Broadband. The goal is to reach the incredible mark of 10 Gigabits/second on the uplink and 20 Gigabits/second on the downlink, allowing you to download a 15 Gigabyte HD video in just 6 seconds. That same video would take 240 seconds (4 minutes) with 4G. With the availability of this service, it will be possible to offer services that are difficult to access today, such as Virtual Reality, Augmented Reality, and even holography, not to mention 3D videos in 4K resolution, surveillance cameras, and online electronic games with a much better performance than those currently available.

The second characteristic is *Latency Reduction*, i.e., the period required to send information to the system and receive a response (2-way), which requires extreme speed. This processing speed is essential for very high-reliability applications. In 5G nomenclature, this factor is called URLLC – Ultra-Reliable Low Latency Communication. This is the case, for example, in applications used in autonomous vehicles and in automated industries, where equipment operating at very high speed such as robotic arms controlled from a distance.

In the specific case of autonomous vehicles, it is necessary to keep the latency time in the order of 1 ms (millisecond). This means that a vehicle forced to stop when it is at 100 km/hour would travel 2.8 cm in this time lapse. Note that this value, 1 ms, refers only to the information transmission time, not to the reacting time. If 4G were employed, the distance would be 1.4 m, i.e., 50 times longer! The same applies to robotic surgery, in which mechanized surgical instruments must obey the doctor's commands instantaneously.

When talking about 1 ms latency, it is always interesting to evaluate what this means in terms of distance, when it comes to propagation of radioelectric signal. Let's remember that the propagation speed of a radio wave is equal to the speed of light, i.e., 300,000 km/s. This means that in 1 ms (10^{-3} s) a wave travels 300 km in free space. One must also consider that after the signal is received, it must be processed before it returns. All this in less than 1 ms? Therefore, any processing that takes place at a distance greater than 300 km, or 150 km if we consider both directions, cannot meet this requirement. It is very important to always keep this value in mind, especially when talking about robotic surgery-type applications where the surgeon is far from the patient.

The third feature is mMTC – Massive Machine Type Communication. With the advent of IoT (Internet of Things) proliferation, the intention is to enable the connection of up to 1,000,000 devices/km^2. If 4G technology were employed, it would be possible to serve only 100,000 devices.

The best graphical representation of the three pillars of 5G is a pyramid, as shown in Fig. 2.1.

5G brings so many innovations that it is difficult to define where to start describing them. But in our view, the starting point is what is known as "Network Slicing." For new services to be provided, it is necessary to identify the nature of each one and treat them differently. To do this, the network must be divided into slices, with each slice assigned performance standards that meet the needs of each service.

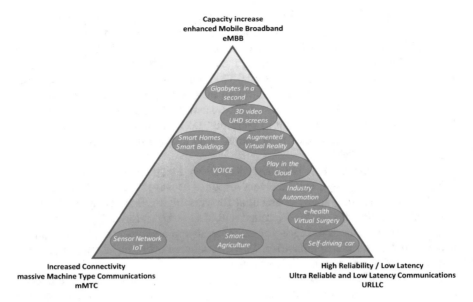

Fig. 2.1 The three pillars of 5G

This slicing will certainly benefit operators that currently make their networks available for content traffic without monetarily participating in the profit of the product that travels over them. This new way of operating networks will give operators the reason they needed to charge according to the content that travels over their networks.

In the vacuum of this evolution, an even more revolutionary function emerges from the point of view of the business model. Until recently, large network equipment manufacturers provided their systems in an integrated manner, i.e., specific hardware associated with dedicated software. This is a complete solution, from installation to operation, practically providing operators with the IP interface ready to be connected to the network and then to routers, which in turn would make the necessary connections to complete the service provided to users. In 5G, the model will be different.

The approach to 5G is revolutionary precisely because the proprietary and exclusive "black box" of each manufacturer is being opened, and each module that makes up the system is being standardized through very well-defined interfaces that will allow interoperability between the various manufacturers of these modules. In other words, the opportunities for new players to participate in the market are becoming enormous.

Another important issue is the insertion of virtualization and cloud techniques in the process. Virtualization allows you to separate hardware and software and obtain important advantages, such as the use of common platforms found on the market or COTS – a Commercial Off-The-Shelf, for hardware implementation, which results in reduced equipment cost. It is possible to install all required applications and network management software in virtual machines (VM), taking advantage of: resource sharing, elasticity, scalability, auto-provisioning, migration between VMs, fault tolerance (FT) among others, which makes the 5G network much more efficient and flexible.

In short, 5G might be an open, non-proprietary system. We can compare the impact of this openness to the revolution that occurred when IBM launched the PC and allowed numerous manufacturers to produce their own motherboards and a wide variety of interface cards, such as for disks, printers, and monitors. This popularized the use of the personal computer, drastically reducing cost and making it an indispensable tool for everyone. 5G uses Open-Source management software and open architectural principles, which did not happen with the PC until the appearance of Linux, which by the way will be of great use in this new context.

Obviously, this did not happen by chance, but by the imposition of the big operators and a multitude of new companies, eager to participate in this billion-dollar market.

From the traditional point of view of commercial access business, the progressive evolution of cloud processing and virtualization will open a huge front for outsourcing services.

This outsourcing process tends to be very intense, like what has already occurred in the automobile industry, which ceased to be a verticalized industry and became a horizontal assembler. Something similar occurred in the electronics segment, in which several companies completely outsourced their production lines and began to devote themselves to research and development of new products.

We should not be surprised if, within a very short period, large operators hand over most of their existing infrastructure to highly specialized companies, which are likely to provide services to many operators without distinction. In this context, operators will be able to focus their efforts on promoting their brands and marketing their products. And they will eventually become new MVNO – Mobile Virtual Network Operators, outsourcing much of the operation of their own networks and making their own structures much cleaner.

It is not difficult to imagine that we will have infrastructure providers, network operators, cloud service providers, application service providers, and network providers specializing in access services. This will allow for greater specialization, cost reduction, and the ability to innovate. Business structures will be cleaner, lighter, and much more efficient. It can also be anticipated that investors will be eager to participate in this new business structure.

Chapter 3
The Evolution of Cellular Technologies

It is always interesting to look at the past to better evaluate the present and the future. It will only be possible to appreciate how much 5G represents in terms of technological evolution if we analyze everything that has occurred since 1980 until today.

First, let's remember the reason behind the development of the transmission system in hexagonal cells, like a beehive. This format is the closest to a circle. The division of an area in cells allows the reuse of frequencies and improves the use of available spectrum, which is really the greatest asset. The power used in BS (Base Station) is sized to meet exactly the desired area limited by the cell and avoid generating interference in cells that reuse the same frequencies used in the cell in question.

With increasing traffic demand, it is often necessary to sectorize a cell, i.e., generally use antennas that radiate with an aperture of 120° (it is also possible to use 60°), allowing the adoption of different frequencies in each sector. This greatly complicates frequency planning, which is almost always done only by specialized software. Figure 3.1 illustrates these comments.

In order to better understand the characteristics of each generation of mobile phones and understand their evolution, it is necessary to characterize some concepts without, however, going into too much technological detail.

Analog Signals Versus Digital Signals

This is a common concept nowadays, but do we all know how the transformation of an analog signal into a digital signal occurs? Well, any analog signal can vary its level discretely and continuously within a certain value range. To be digitized, these signals go through a process called quantization (evaluation of the analog signal level and generation of a binary code that represents the corresponding analog level). The appropriate sampling frequency must be maintained so that the various levels are transformed into binary codes (0 or 1) with compatible periodicity. The

J. L. Frauendorf, É. Almeida de Souza, *The Architectural and Technological Revolution of 5G*, https://doi.org/10.1007/978-3-031-10650-7_3

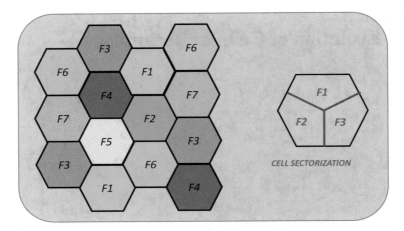

Fig. 3.1 Frequency planning

sampling frequency is an important value so that there is no loss of information. From these binary values, the signals can be processed more simply and are more easily handled by logic circuits and microprocessors.

Carrier Wave

By definition: *"Waves are disturbances that propagate from one place to another through a medium or even in a vacuum, transporting energy. They are disturbances that move in space transporting energy exclusively from one point to another, without transporting matter".*

Fig. 3.2 Wave characteristics

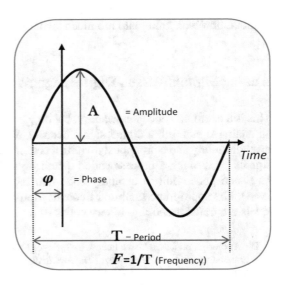

These are the waves that allow the propagation of radioelectric radiations along the entire Radiofrequency Spectrum. The main characteristic of a wave is its frequency (F), but two other points are also important: its amplitude (A) and its phase (φ). It is exactly these three quantities, shown in Fig. 3.2, that allow us, through their changes, to send information. In this case, the wave is called a *carrier* wave.

By varying the amplitude, frequency, or phase of the signal, it is possible, through changes, to transmit information in digital systems, as shown in Fig. 3.3. This is how telecommunications technology has evolved and is transforming our lives.

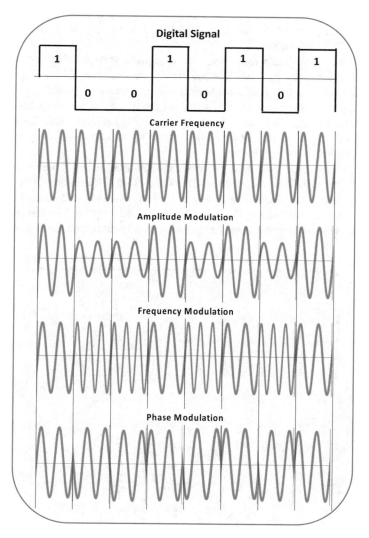

Fig. 3.3 Possible ways of modifying a radio signal

Digital Modulation

It is possible to combine the amplitude and phase variations, keeping the frequency unchanged in such a way to obtain, through these combinations, different binary codes, which generate, in turn, different bits of information. Just by varying the phase, for example, it is possible to generate two symbols that represent 1 bit of information, as shown in Fig. 3.4.

In the figure, the transition occurs exactly when the sinusoid goes from the upper quadrant to the lower quadrant (0° ==> 180° or 180° ==> 0°). If the phase transition occurs in other parts of the sinusoid, which can be at 90°/180°/270°/360°, other binary codes can be created.

In addition to the phase transitions of the signals, it is possible to change their amplitude simultaneously. Thus, by combining the two variables, it is possible to change the carrier so that it allows the transport of a greater volume of information (bits). The greater the number of bits that a carrier can carry, the greater the efficiency of the process. This index is called *Spectral Efficiency*. Table 3.1 shows the main forms of digital modulation used in cellular systems and their respective spectral efficiencies.

It is important to consider that, the lower the modulation level, the more robust is the signal, because it is easier to discriminate what information it is carrying. Explaining: when the signal propagates, it suffers attenuation and interferences. If this attenuation is too intense, the signal may not be recognized by the receiver. In addition, there is always noise in the reception generated by the receiver itself. It is

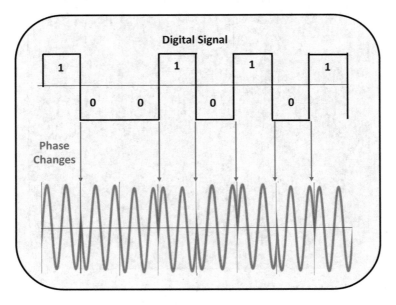

Fig. 3.4 BPSK – Binary Phase Shift Keying digital modulation

Table 3.1 Main parameters of digital modulation

Type of modulation	Spectral efficiency (bps/Hz)	Carrier level/noise level (C/N ratio dB)
PSK	1	10.6
QPSK	2	13.6
4-QAM	2	13.6
8-QAM	3	17.6
8-PSK	3	18.5
16-PSK	4	24.3
16-QAM	4	20.5
32-QAM	5	24.4
64-QAM	6	26.6

C/N Ratio (CNR) stands for carrier-to-noise ratio and is measured after modulation
S/N Ratio (SNR) stands for signal-to-noise ratio and is measured before modulation

important to define a metric for how to evaluate the quality of reception. The metric used for this purpose is a ratio identified as *C/N* (***Carrier/Noise Ratio***). It tells us how much the carrier level (*Carrier*) is above the noise level (*Noise*). Because it is a logarithmic scale, this ratio is expressed in dB (Decibel). It can be assumed that in the most striking situations, like those shown in Fig. 3.4 (0° ==> 180° or 180° ==> 0°), it is easier to identify the transition from one phase to another. As other points on the sinusoid are used to inform the transition, it becomes more difficult to make this identification (discrimination) and, therefore, the signal that is received must be at higher levels relative to the noise.

The rightmost column of Table 3.1 gives an idea of minimum values for the C/N ratio that allows the type of modulation to be used. You can imagine that the closer the user is to the base station, the better the modulation level and the more efficient the transmission and quality of the connection. Many systems, as we will see later, use this relationship and adjust the modulation level to the existing conditions. This evaluation is dynamic and can vary throughout a transmission. The evolution of the various generations of cellular systems is due precisely to the advance of technology that allows us to use more complex types of modulation, thanks to the ability of electronic circuits to have sufficient sensitivity to separate the received signals from the noise of the reception.

Transmissions also use parity codes like those used in most numeric transactions, such as banking, which require long strings of numbers. These parity codes allow you to identify whether the message has arrived intact or not. If not, the system requests that the message be retransmitted. The transmission error rate is called *BER – Bit Error Rate*. It is exactly the BER that identifies the quality of the link received.

Another method used is sending two signals with the same content by different paths or carriers. When the signals are received by one or more independent paths, an analysis is made by special mathematical algorithms that work by employing a methodology of signal correlation. This makes it possible to separate information, which is a deterministic quantity, from noise, which is probabilistic in nature. We will see that this technique is widely used in 5G. Needless to say, this has only become possible due to the enormous advances in processors, which can process information in fractions of a second.

Frequency Spectrum

Radioelectric waves are part of the radioelectric spectrum. This spectrum is used to transmit radio waves, such as AM radios, which use amplitude variations for information transmission; FM radios, which use frequency variations for their operation; and even TV stations, which used frequency variations for sound transmission and amplitude for video transmission when they were still analog. All these services occupy the initial part of the spectrum, as shown in Fig. 3.5.

Cellular services initially occupied bands from 800 to 900 MHz (UHF – Ultra-high Frequency) or even lower frequencies according to the availability of spectrum. As they evolved, systems had to occupy higher bands, SHF (Super High Frequency) and EHF (Extremely High Frequency), and the reason is simple: as the amount of information to be transmitted grows, so does the need for greater bandwidth. Larger bandwidths are only available at significantly higher carrier frequencies. This is exactly why 5G systems will occupy much higher frequencies than those used today, frequencies in the 3.5 GHz (SHF), even 28 GHz, 60 GHz (EHF) or even higher frequencies. 4G operates in the 2.6 GHz range, while previous generations did not go beyond 2.1 GHz. Also remember that Wi-Fi operates in the 2.4 GHz and 5 GHz bands (unlicensed spectrums).

Just as an example, the first mobile phones occupied frequency bands of 25 + 25 MHz (824–849 MHz and 869–894 MHz). In the second generation of mobile phones, the frequency bandwidth was extended to 60 + 60 MHz (1920–1980 MHz and 2110–2165 MHz + 1895–1900 MHz) reaching up to 75 + 75 MHz (1710–1785 MHz and 1805–1880 MHz), which corresponds to the 1.9 GHz and 1.8 GHz licensed bands. 4G operating at 2.6 GHz occupies a bandwidth of 190 MHz (2500–2690 MHz). For 5G, the bands will be even larger, for

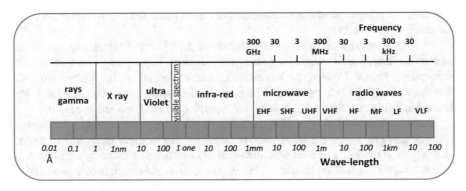

Fig. 3.5 Radio frequency spectrum

example 400 MHz, considering the frequency range from 3.3 to 3.7 GHz. For higher frequency bands, we have values like 24.3–27.5 GHz. The bands will be measured in Giga Hertz (GHz) and no longer in Mega Hertz (MHz).

Low bands, such as the 700 MHz (UHF) band, will also be useful for 5G to serve special applications where coverage is more important than the volume of information to be transmitted. It is important to remember that the attenuation of radiations in free space (FSPL) is directly proportional to the square of the transmission frequency.[1] So, the higher the frequency, the smaller the coverage area of a cell.

But in order to clarify the terms we are using, it is important to characterize them:

- *Operation Band* is the portion of the spectrum reserved by an operator and within which it must operate.
- The *Carrier* Frequency is the frequency that carries the information.
- *Operating Channel (Band)* is the portion of the spectrum that the carrier can occupy when modulated for the transmission of the information to be sent.
- *Spectral Efficiency* is the index used in digital systems to measure how effectively a technology uses the available spectrum resources. The measurement parameter is equivalent to the number of bits (amount of information) that can be transmitted in a 1 Hz.

The word multiplexing will appear dozens, hundreds of times in the next few paragraphs. *Multiplexing* is the way to send multiple simultaneous communications/information using a single carrier.

The way to use the spectrum has also evolved and this will become very clear in the next paragraphs. There are basically two forms of spectrum partition to enable bidirectional transmission, (UE) user ==> (BS) base station (uplink) and in the opposite direction BS ==> UE (downlink), which are the FDD (Frequency Division Duplex) and TDD (Time Division Duplex). Figure 3.6 shows the difference between the two schemes used.

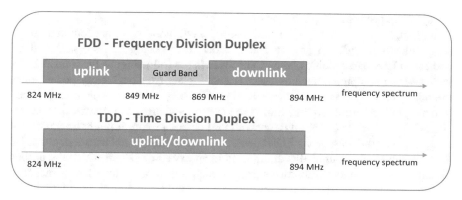

Fig. 3.6 Ways of dividing the frequency spectrum

---

[1] **FSPL** (Free Space Path Loss) = $(4\pi df/c)^2$; where, f = signal frequency; d = distance; c = speed of light)

Let's use as an example the first spectrum allocated by operators who used the 850 MHz band to put 1G in operation in their networks. In those first networks, the spectrum was divided into two bands, the lower one, from 824 to 849 MHZ, and the upper one, from 869 to 894 MHz. Between the two bands was left an unused band called Guard Band. This way of dividing the spectrum is called *FDD* – Frequency Division Duplex and allocates a portion of the spectrum for the communication of the user' device with the Base Station (BS), that we will adopt as the up/uplink direction, and another portion of the spectrum for the opposite direction, i.e., communication of the BS with the users' device, that we will adopt as the down/downlink direction. The Guard Band is left to prevent interference between the different directions of communication. This form of spectrum allocation is a legacy of the format used to transmit voice, since the content in both directions has the same bandwidth (in this case, 25 MHz). This does not apply to data communications where, in general, the downlink bandwidth is much larger than the uplink bandwidth.

It would be more logical, in the case of digital systems in which data communications predominate, that the spectrum is shared differently, not reserving separate bands for each of the two directions, but using the entire available spectrum, without the loss of the Guard Band, and sharing this spectrum bidirectionally, so that one moment it is used by the uplink and the next by the downlink. This way of sharing spectrum is called *TDD* – Time Division Duplex. Of course, this switching in the direction of transmission is done very quickly, imperceptible in communication, especially for services such as voice. It would be like systems such as radio communication in which each user must say "over" to allow the interlocutor to respond, obviously making this switching of the senses very fast. This system increases the spectral efficiency because it allows dynamically allocating more or less time for transmission in each direction, depending on the volume of data flowing in each, besides not using Guard Band, which in the example of the illustration would represent 40% of gain! In addition, as both directions use the same transmission channel, the system is constantly monitoring the propagation conditions, being much easier to adjust the propagation parameters optimally.

Although perhaps it is not the time, but to leave no doubt, we are forced to make a parenthesis in order to prevent some reader from raising an issue that will be addressed later in a specific chapter. The issue is the following: as the transmission power of the equipment used by users is much lower (to save battery power) than the power used by BSs, the communication efficiency is lower in the uplink direction and, as a result, it requires more time to send a volume of information like the corresponding volume sent on the downlink. For this reason, it might be more reasonable to continue using the FDD format. In practice this is not true and, even if it were, because of dynamic allocation, even in an extreme case, in which the uplink direction requires more time for its transmission, this would also be possible. It is important to note that the time partition must be valid for an entire region managed by the system that monitors the traffic volume in both directions. The synchronism of the entire network is essential and constitutes one of the most critical requirements of all these new technologies and, in particular, of TDD networks.

Another important concept is to analyze how the available spectrum can be divided to optimally accommodate the largest number of users to be served simultaneously. This process, as defined above, is the multiplexing of content generated by multiple users.

There are several ways of multiplexing the channels:

- *FDMA* – Frequency Division Multiple Access, i.e., slicing the frequency spectrum by dividing it into small bands. Each small band is allocated to a user or a carrier.
- *TDMA* – Time Division Multiple Access consists of allocating to each user or carrier a fraction of the time reserved for the channel, which is called *time slot*. Note that this has nothing to do with the TDD/FDD scheme. TDMA works both in TDD and FDD. TDMA is a form of multiplexing communication channel.
- *CDMA* – Code Division Multiple Access consists of transmitting the content of several users simultaneously, with each user receiving a specific code, using a process called Spread Spectrum.
- *TD-CDMA* – Time Division CDMA, which uses the CDMA system, but simultaneously slices the transmission period into time intervals, as is done in the case of TDMA.

Figure 3.7 graphically shows the four ways to multiplex the content of services to be processed simultaneously.

Having learned these concepts, let's evaluate how cellular networks have evolved over time. Figure 3.8 shows the evolution of the various generations.

At a first glance, we can observe how difficult it was to convince operators from different continents to converge to a single standard. Imagine the problems faced by mobile phone manufacturers to launch equipment that would allow them to operate, even within the same technology, and that would work anywhere in the world in order to guarantee easy roaming. It took 30 years for this to become a reality and another 10 years for real convergence to occur. Only with the arrival of the fourth generation, with the advent of LTE, did standards effectively converge, but still with different operating dynamics, just because while most operators insisted on maintaining the division of the spectrum according to the FDD – Frequency Division Duplex standard, others opted for TDD – Time Division Duplex.

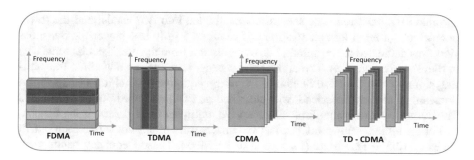

Fig. 3.7 Ways of dividing the spectrum in multiplexed communication channels

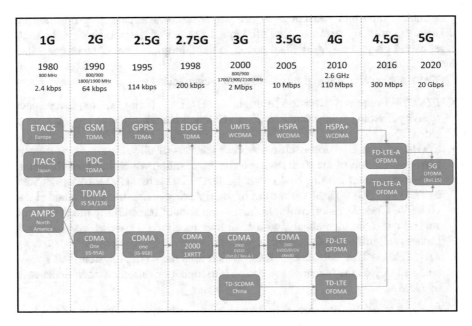

Fig. 3.8 Generations of cellular systems

The Chinese migrated to LTE, but chose the TDD scheme, as they had been doing since 3G, so as not to divide the spectrum into distinct bands for uplink and downlink, but rather to use the entire band for two-way communications. With the advent of 4G this problem was solved, not by changing the concept of spectrum partitioning, but by the ability of the system to process both schemes. It was also with 4G that TDMA and CDMA modulation systems were replaced by a new concept, OFDM. In the next chapter we will discuss in detail what is OFDM.

Let's learn more about and analyze each of the various generations. Figure 3.9 summarizes the evolution of the available services.

The First Generation – 1G – was launched in 1980. It was called AMPS (Advanced Mobile Phone Service/System) in the United States. It was an analog system that used frequency modulation (FM), the same used in analog television and radio transmission. The spectrum was divided into two bands; and the lower one, which suffered less attenuation and required a little less power for transmission, was dedicated to the uplink, i.e., transmission from the user's equipment (UE) to the Base Station (BS). The upper band was dedicated to the downlink, transmitting signals from the Base Station to the user's equipment. This form of spectrum division, as already described, was classified as FDD and used 30 kHz channels. The frequency division multiplexing between multiple access (users) is the FDMA.

For the uplink + downlink bands, two carriers were allocated, and to each one fit slices of 12.5 + 12.5 MHz (for a total of 25 MHz). These frequency bands were divided into 30 kHz channels (where each channel consists of a pair of frequencies

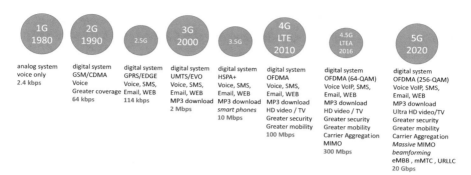

Fig. 3.9 Evolution of cellular systems – main characteristics

used in Transmission and Reception), which in the case of AMPS system allowed, therefore, to have up to 416 individual channels. Of these 416 channels, 395 allowed simultaneous conversation of users, and the remaining 21 channels were used to transmit control information using a system called FSK (Frequency Shift Keying), which allowed to transmit information at a rate of 2.4–10 kbps.

Europe adopted a similar system, ETACS (European Total Access Communication System) with only a few different parameters, such as channel width of 25 kHz instead of 30 kHz, which allowed operators to have a larger number of simultaneous users, up to 500. The European system was not unique, having variants in Scandinavian countries, Germany, France, and England. Despite this, in 1991, the world had ten million connected devices.

Following the success of mobile telephony, two distinct technological strands occurred, one following the TDMA technology (European line) and the other, the CDMA line (North American line). It makes more sense to follow the evolution of each of these strands until their confluence, which only occurred 20 years later with the arrival of 4G.

Let's chase the European line first, which turned out to be the most successful due to the largest number of users.

European Evolution – TDMA: 2G and 3G

The European Second Generation – 2G – launched in 1991 in Finland was the first generation to use *digital voice* in mobile systems. Fixed telephony already employed voice digitization for long-distance transmission (PCM – Pulse Code Modulation), although fixed-line handsets remained analog, as is the case in most cases to this day.

The main system employed in Europe to replace ETACS was GSM (Global System for Mobile – GSM1xRTT), employing TDMA. Due to its great success, several frequency bands were used, 890–915 MHz for uplink and 935–960 MHz for downlink, and 1710–1785 MHz for uplink, and 1805–1880 MHz for downlink. The carriers each occupied 200 kHz of spectrum. Each carrier could serve up to eight simultaneous users with voice transmission rate of 9.6 kbps. So, $9.6 \times 8 = 76.8$ kbps at 200 kHz, which is equivalent to a Spectral Efficiency of 0.38 bps/Hz for voice transmission. Used the carrier modulation system called GMSK (Gaussian Minimum Shift Keying) in this first stage. Maintaining the band of 12.5 MHz for each carrier, it was possible to attend 496 users, i.e., 992 users in total, approximately the same quantity as the analog system, but with quality and privacy gain of the conversations. In addition, it was possible to transmit data using the circuit-switched system at a rate of 9.6 kbps. Theoretically, it is possible to reach transmission rates of up to 64 kbps using the 8 slots available in one carrier. GSM employed a high-speed switching system, the HSCSD (High Speed Circuit Switched Data). These parameters refer to GSM Rev.1.

We observed, however, that in 2G data followed the same path as voice service and depended on the traditional system used in telephone exchanges, the Circuit Switched system.

This new generation brought with it the possibility of sending text messages, SMS (Short Message Service), which is used until today. Another great innovation was the use of a chip to identify the user, the SIM Card (Subscriber Identity Module), which brought greater security to the system. Another advantage of GSM was allowing roaming between more than 200 different countries. In fact, the convenience of GSM roaming has been one of the key factors for the global success of the platform, which in this first phase alone had 90 million users.

2.5G Generation The 2G, however, needed to provide an internet access service and soon received an update, an upgrade, and became known as 2.5G. The main innovation was the introduction of *packet switching* (PS – Packet Switched), the same transmission technique adopted by IP networks (TCP/IP – Transmission Control Protocol/Internet Protocol architecture). The 2.5G network was connected to a packet data network (PDN), but the connection to the *circuit switched network* was maintained for the voice channels. The separation of the voice service from the data service allowed for greater efficiency in processing the data service. This innovation received the acronym *GPRS* (General Packet Radio Service). This was the revision 5 (GSM Rev.5), which allowed reaching rates of 115.2 kbps by using the 8 available slots ($8 \times 14.4 = 115.2$ kbps). With this, the spectral efficiency increased to 0.58 bps/Hz. In very favorable conditions, it would be possible to reach up to 20 kbps/slot, which would allow reaching up to 160 kbps. Typical implementations, however, were in the order of 32–48 kbps for the downlink and 15 kbps for the uplink.

2.75G Generation The 2.5G GPRS also received another innovation that allowed, thanks to a new carrier modulation system (8-PSK = 8 states Phase Shift Keying) and different error correction, increase the spectral efficiency. While GSM/GPRS processed 1 bit/symbol with GMSK modulation, with this new 8-PSK modulation was able to process up to 3 bits/symbol, practically tripling the volume of transmitted data and receiving the name *EDGE*. It was then possible to reach up to 59.2 bps/

slot. Using the eight available slots, it was theoretically possible to reach 473.6 kbps. This innovation, however, meant that users had to change their handsets.

EDGE also predicted an EDGE-Evolution, with Revision 7. The goal was to maintain the acquired leadership. To do so, it foresaw the use of higher modulation levels, 16-QAM and 32-QAM, besides other innovations, to reach the level of 1.9 Mbps, always considering the downlink rate. But the evolution happened differently, as we will see below.

Anyway, this second generation of GSM and its successors, GPRS and EDGE, in 2011, according to the Informa Telecoms and Media bulletin, reached 4.2 billion users on its networks spread across 220 countries of the globe.

Third European Generation – 3G The Europeans abandoned, in its third generation, TDMA and switched to CDMA used in North America for 2G. But the European implementation adopts a new concept, the Wideband version, that uses 5 MHz bands instead of the 1.5 MHz band used by the North American CDMA, as we'll see later. The UMTS – Universal Mobile Telecommunications Service was launched using W-CDMA – Wideband CDMA. With this, it was possible to reach speeds of 2 Mbps. The technology allowed its use in both FDD and TDD implementation, although most operators adopted the FDD version.

This new technology served up to 100 simultaneous voice users with a 5 MHz bandwidth or allowed data transmission that could reach peak rates of 384–2048 kbps on the downlink. It also brought two new features, support for adaptive modulation for each user, which allowed to increase the transmission rate according to individual conditions.

It is interesting to say that the "European" system, having been standardized by IMT (International Mobile Telecommunications) and having been baptized as IMT-2000, was obviously not a European exclusivity, so much so that it was initially launched in Japan by NTT DoCoMo in 2001. It turned out that GSM became a recognized international standard, and in 1998, six international telecommunications regulatory institutions joined together to define a single standard, UMTS, which became the third generation of cellular systems. The first version of UMTS was launched in 1999 and, for this very reason, was given the name UMTS Release 99.

3.5G Generation The technology moves on to more ambitious goals. UMTS evolves to HSPA (HSDPA – High Speed Downlink Packet Access + HSUPA – High Speed Uplink Packet Access), reaching the 10 Mbps mark.

HSDPA increased the transmission rate on the downlink using a technique called AMC (Adaptive Modulation and Coding), which, according to the channel conditions, allows modulation levels ranging from QPSK (Quadrature Phase Shift Keying) to 16-QAM (Quadrature Amplitude Modulation). With these improvements, it was possible to reach rates from 10 to 14 Mbps. In this stage, the uplink remained unchanged, with speed around 384 kbps.

Table 3.2 Comparison of data transmission speeds

TECHNOLOGY	DOWNLINK			UPLINK		
	Network Peak Value	User Peak Value	User Typical Values	Network Peak Value	User Peak Value	User Typical Values
EDGE - Type 2 MS (2 x frequency channels / Mobile Station)	473.6 kbps			473.6 kbps		
EDGE - Type 1 MS Practical Terminal (1 x frequency channels / Mobile Station)	236.8 kbps	200 kbps	160-200 kbps (4/5 timeslot = 4/5x40 kbps)	236.8 kbps	200 kbps	80-160 kbps (2/4 timeslot = 2/4x40 kbps)
Evolved EDGE (type 1MS) (receive up-to-ten timeslots using two radio channels / transmit up-to-four timeslots in one radio channel using 32-QAM)	1,184 kbps (10 slots/dual carrier = 10x118.4 kbps)	1 Mbps	350-700 kbps(Dual carrier)	473.6 kbps (4 timeslots = 4x118.4 kbps) (Dual carrier)	400 kbps (4 timeslots = 4x118.4 kbps)	150-300 kbps (2/4 timeslots = 2/4x40 kbps) (Dual Carrier)
EvolvedEDGE (type 2MS). (receive up-to-six timeslots using two radio channels/ transmit up-to-eight timeslots in one radio channel using 32-QAM)	1,894.4 kbps (16 slots/dual carrier = 16x118.4 kbps)			947.2 kbps (8 timeslots = 8x118.4 kbps) (Dual carrier)		
Evolved EDGE (16 carriers) (16 carriers for downlink, 200 kbps/carrierr using 8 PSK / MIMO)	6.4 Mbps					
UMTS WCDMA Release 99	2,048 kbps			768 kbps		
UMTS WCDMA Release 99 Pratical Terminal	384 kbps	350 kbps	200-300 kbps	384 kbps	350 kbps	200-300 kbps
HSDPA Initial Device (2006)	1.8 Mbps	>1 Mbps		384 kbps	350 kbps	
HSDPA	14.4 Mbps			384 kbps		
HSPA Initial Implementation (HSPDA + HSUPA)	7.2 Mbps	>5 Mbps	0.7 - 1.7 Mbps	2 Mbps	>1.5 Mbps	0.5 - 1.2 Mbps
HSPA	14.4 Mbps			5.76 Mbps		
HSPA+ (DL 64-QAM / UL 16-QAM / 5+5 MHz)	21.6 Mbps		1.9 - 8.8 Mbps	11.5 Mbps		1 - 4 Mbps
HSPA+ MIMO (2T2R / DL 16-QAM / UL 16-QAM / 5+5 MHz)	28 Mbps			11.5 Mbps		
HSPA+ MIMO (2T2R / DL 64-QAM / UL 16-QAM / 5+5-MHz)	42 Mbps			11.5Mbps		
HSPA+ (DL 64-QAM / UL 16-QAM / Dual Carrier 10+5 MHz)	42Mbps		3.8 - 17.6 Mbps (doubling 5 MHz)	11.5 Mbps		1 - 4 Mbps
HSPA+ MIMO (2T2R / DL 64-QAM / UL 16-QAM / Dual Carrier 10+10 MHz)	84 Mbps			23 Mbps		
HSPA+ MIMO (2T2R / DL 64-QAM / UL 16-QAM / Quad Carrier 20+10 MHz)	168 Mbps			23 Mbps		
HSPA+ MIMO (2T2R / DL 64-QAM / UL 64-QAM / 8 Carrier Down / Dual Carriers Up (40+10 MHz)	336 Mbps			69 Mbps		
LTE MIMO (Single user) (2T2R / 10+10 MHz) (Network performance)	70 Mbps		6.5 -26.3 Mbps 5-12 Mbps	35 Mbps(64-QAM) 22 Mbps(16-QAM)		6.0 - 13.0 Mbps 2-5 Mbps
LTE MIMO (4T4R / 20+20 MHz)	300 Mbps			71 Mbps(64-QAM) 45 Mbps(16-QAM)		
LTE-A MIMO (8T8R / DL 64-QAM / UL 64-QAM / 20+20 MHz)	1.2 Gbps			568 Mbps		
CDMA2000 1XRTT	153 kbps	130 kbps		153 kbps	130 kbps	
CDMA2000 1XRTT	307 kbps			307 kbps		
CDMA2000 EV-DO Rel 0	2.4 Mbps	>1 Mbps		153 kbps	150 kbps	
CDMA2000 EV-DO Rev A	3.1 Mbps	>1.5 Mbps	0.6-1.4 Mbps	1.8 Mbps	>1 Mbps	300-500 kbps
CDMA2000 EV-DO Rev B. (3 radio channels/5+5 MHz)	14.7 Mbps			5.4 Mbps		
CDMA2000 EV-DO Rev B / Theoretical (15 radio channels / 20+20 MHz)	73.5 Mbps			27 Mbps		

PICO values = maximum allowed by the technology
TYPICAL Values = values obtained when sending FTP (File Transfer Protocol) type files
Data taken from Mobile Broadband Explosion, Rysavy Research/4G Americas, August 2013

HSUPA, from the other side, increases the uplink transmission rate using a dedicated transport channel called E-DCH (Enhanced Dedicated Channel) and uses the same air interface improvement resources that were employed for the HSDPA version. Thus, it was possible to reach uplink rates of around 5.76 Mbps.

The latest evolution of this technology was named HSPA+ and incorporated a series of innovations, such as MIMO (Multiple Input Multiple Output) antennas, in which two transmit and two receive antennas (2T2R) are used at this stage of the technology, using an algorithm developed by Siavash Alamouti called Space-Time Coding, which employs both spatial and temporal diversity. In addition, he added 64-QAM as a new modulation alternative for the downlink carrier and 16-QAM for the uplink one. In this step, can also be used the technique of aggregating up to two carriers (2×5 MHz = 10 MHZ – Dual Carriers), allowing reaching the peak rate of up to 84 Mbps in the downlink, using 64-QAM and MIMO. Adding 4 carriers (20 MHz), this rate increased to 168 Mbps or, adding 8 carriers (40 MHz), it was possible to get 336 Mbps. Obviously, these values are all theoretical. Table 3.2 shows all stages of this evolution. The highlighted lines indicate actual measured values and not theoretical ones.

North American Evolution – 2G and 3G

In North America, with the digitalization of AMPS, cellular telephony evolved to Digital AMPS, or *D-AMPS,* or *TDMA (IS-54/136).*

IS-54 (1980) used TDMA multiplexing. Both 1G bands, 824–849 MHz for the uplink, and the 869–894 MHz band for the downlink, and the 1850–1910 MHz frequency band for the uplink and the 1930–1990 MHz band for the downlink were used. Again, the 20 MHz was kept as the Guard Band. IS-54 followed the same channel format as AMPS, using 30 kHz channels, which were divided into three Timeslots (10 kHz), allocating a Timeslot to each user, which allowed tripling the number of users in the network. The use of cryptography guaranteed the privacy of communications, which did not occur with AMPS. The use of DQPSK (Differential Quadrature Phase Shift Keying) modulation allowed reaching rates of 48.6 kbps, which gave a spectral efficiency of 1.62 bps/Hz. This represents a 20% increase over the efficiency of the competing CDMA system, which will be analyzed next. However, this difference did not make it more attractive than CDMA as it demanded more power from handsets, which reduced battery usage time.

The *IS-136* is an evolution of the IS-54 and kept the same basic features but was launched with the aim of including some advances, such as the possibility of sending SMS text messages and allow its use for data transmission using, for this, the CSD – Circuit Switched Data. With the improvement of voice channel compression and codification, it was possible to double the capacity of each original Timeslot, thus sextupling the capacity of voice channels when compared to AMPS.

This technology evolved to the European system, GSM/GPRS/EDGE, already described in the previous paragraphs.

CDMA-One/IS-95 At the same time (1989), a previously unknown company based in San Diego, California, called QUALCOMM proposed a different modulation technology to replace AMPS, employing CDMA (Code Division Multiple Access) as an alternative to TDMA. The interesting thing about this technology is that all users simultaneously use the same channel, with each user assigned a different code. This makes all cells able to use the same frequencies, eliminating the problem of interference between cells and facilitating the operation. It also adopts a mechanism called "Breathing," which allows a cell to transfer to a neighboring cell part of the traffic when the first cell begins to be saturated, changing the geographic position of the area served. With this, any moving user is also transferred from one cell to another in a much simpler way, called Soft-Handoff. Thus, a call is only disconnected from a Base Station when it is already connected to the new BS. It also uses a process of "voice activity detection," which keeps the terminals inactive while they are not transmitting, saving battery power and reducing interference, which increases system capacity. In this way, CDMA was able to increase the capacity of the voice service when compared to GSM. With the operation of the systems, it was possible to demonstrate the better performance of CDMA in terms of coverage and capacity when compared with GSM. This favored the adoption of CDMA as the system to be used in the third generation of GSM/GPRS/EDGE.

CDMA kept the spectrum partition used by AMPS, i.e., FDD scheme, using 824–849 MHz band for the uplink, and the 869–894 MHz band for the downlink, with Guard Band of 20 MHz. The channel width, in this case, was 1.25 MHz, allowing for 20 carriers, each housing 64 voice channels at a rate of 9.2 kbps or less, totaling 1280 simultaneous voice channels. In addition to the advances in the number of voice channels, the IS-95A version also supported dedicated data channels at a rate of 9.6 kbps. In IS-95B, the Packet Switched System was introduced, replacing the Circuit Switched System, and improved data transmission efficiency (SCH – Supplemental Code Channel), by combining 7 SCHs and one traffic channel to support rates of 14.4 kbps per channel, thus achieving 115.2 kbps.

The increase in transmission speed allowed the evolution of SMS to MMS (Multimedia Message Service) and the consequent sending of images such as photos, graphics, and maps. Thus, mobile phones became multimedia terminals that allowed the instant sharing of information. This became popular with 3G.

Both in North America and in Europe, the evolution from 1G to 2G allowed a significant increase in the quality and quantity of voice services and the beginning of data transmission services with the adoption of Packet Switched, breaking a technological paradigm as it began to replace Circuit Switched. This was the first step towards systems that work exclusively with data. But these new services could not meet the demand for higher transmission speeds. A new alternative became necessary.

While GSM progressed, becoming GPRS and finally EDGE, CDMA evolved to become the definitive standard for 3G cellular telephony, both for Europeans, who had initially adopted it as the standard for 2G – TDMA, and for the American CDMA side, which had already chosen it from the beginning.

The ITU (International Telecommunication Union) gathered in 1999 operators and equipment providers to define the guidelines for the new generation of cellular systems. The group in charge was called 3GPP (Third Generation Partnership Project). Although most of the CDMA development work had been carried out by QUALCOMM and its partner companies, the standardization of this new generation of mobile phones was handed over to this group. Unsurprisingly, the CDMA community was able to evolve to 3G much more easily than the GSM/GPRS/EDGE side. CDMA allowed to fulfill the 3G requirements without having to change the 1.25 MHz channelization and, in this way, it was possible to keep compatibility with the previous systems, which made the migration much easier for its operators, which evolved quickly to CDMA200-1X and EV-DO systems, as we'll see below.

The Third North American Generation Europeans and North Americans converged, therefore, about the form of multiplexing the channels, but not about the width of the channels. North Americans kept the channelization of 1.25 MHz, ratifying the compatibility with the previous system IS-95, while Europeans went for wideband with 5 MHz of channel width. With this, CDMA2000-1X RTT became the first step in the evolution of IS-95. The 1X indicates the use of the same carrier of IS-95 and RTT (Radio Transmission Technology). This version, as already reported, keeps the channelization, but adds 64 new logical channels, which were called Supplemental Channels to increase the traffic capacity of the system. This was only possible by making these new channels "orthogonal" to the existing ones. The orthogonality of carriers is a technique that allows increasing the number of carriers so that they do not interfere with each other. This technique was also used in 4G and 5G technologies, as we will see later.

Even with the addition of these logical channels operating at rates of 9.6 kbps and the use of multiple supplementary channels, it was only capable of reaching 307 kbps and, for that reason, this technology was considered 2.75G and not 3G. The 3G standard stipulates speeds of 2 Mbps for stationary service within buildings, 384 kbps for users who are walking, and 144 kbps for users who are moving at greater speed in cars or trains. In addition, it enables voice services, e-mail access, Web browsing, multimedia streaming, and interactive games.

Through carrier aggregation, it was possible to get close to 2 Mbps, as in CDMA2000-3X. In this case, the band used would be 3×1.25 MHz = 3.75 MHz, getting close to the European band of 5 MHz

The CDMA2000-1X system nearly doubled the capacity of IS-95 by adding the 64 channels dedicated to traffic, orthogonal to the original 64-channel set. It also initiated the use of advanced antennas, which allowed transmission path diversity to be exploited. IS-95A and IS-95B were fully compatible and could use the same carriers, facilitating migration.

But CDMA2000-1X evolved, reaching 2.4 Mbps (downlink) with the CDMA 1XEV-DO (Release 0) system. The acronym "EV-DO," which initially stood for Evolution Data-Only, later changed to Evolution Data-Optimized, so that it would not be mischaracterized. As the name implies, there was no evolution in the voice channels, but only in the data channels that used exclusive carriers for this purpose. The system has then an asymmetric characteristic, since the uplink continues with a speed of 153 kbps (Release 0) and uses BPSK modulation. Typical speeds were only between 300 and 600 kbps (downlink) and 70 and 90 kbps (uplink) (Release 0). CDMA 1XEV-DO had to go through several upgrades in search of better performance.

EVDO used TDMA, splitting CDMA carriers into time slots. It had then to allocate specific carriers for data service. Besides that, it started to adopt 16-QAM modulation, as an alternative to QPSK.

Developed as a High Data Rate (HDR) solution, initially the EVDO standard was well suited to meet the fixed and nomadic services, in which the user can change position within the cell, as the service is provided in a stationary way, not meeting the mobility requirements. The fact was quickly overcomes. In 2002, 3G-1X EV-DO already had the technology to fully meet 3GPP's requirements, delivering mobile internet at 2.4 Mbps to the end user, something that HSDPA could only meet in 2005, 3 years later. In July 2009, according to data from the CDMA community, EV-DO reached 120 million users.

To evolve, EV-DO had to use TDMA to divide into time slots the CDMA carriers, becoming a TD-CDMA system (downlink). Moreover, it started to use three levels of modulation in its carriers: QPSK, 8-PSK, and 16-QAM. Depending on propagation conditions, downlink speeds could vary from 38.4 to 2457.6 kbps, thanks to adaptive modulation technology. That was Revision 0.

In the next revision, CDMA 1XEV-DV Rev.A (Evolution Data and Voice Optimized), the symmetry was improved, allowing it to achieve 1.8 Mbps on the uplink, while the downlink was also improved, reaching 3.07 Mbps. With this, the carriers could be used for both data and voice through a technology like that used in HSDPA. In commercial applications, 450–800 kbps for the downlink and 300–400 kbps for the uplink were achieved.

With Revision B (2008), EVDO/EVDV, using multiple carriers, reached speeds of 14.7 Mbps (64-QAM) in the downlink and 5.4 Mbps (16-QAM) in the uplink. With 3 carriers, started to occupy the band of 3.75 MHz, comparing to the European system, with 5 MHz.

The evolution of cellular technology has been incredible, although the number of versions, which on average lasted less than 5 years, has cost operators a lot. But finally, the world began to converge and reached the fourth generation, which had WiMAX as the first technology and was a disruptive element on the world stage.

However, it was not accepted by operators, who opted for LTE. In fact, as we will see later, LTE/4G and its precursor WiMAX were the watershed. It was only at this point that telecommunication, driven by cellular technology, came of age.

With all the spotlight on the convergence of technologies, the fundamental and necessary synergy was created for the rapid progress from 4G to 4.5G and finally to 5G.

Chapter 4
Digital Modulation Used in 4G/LTE

Up to the previous chapter, we have presented information and technical concepts in general, seeking to provide an overview of the main points of the evolution of mobile telephony and enabling the global context to be understood as fully as possible. As we are now going to talk about 4G and 5G, we are obliged to delve into the concepts that underpin these technologies and that have enabled the latest advances. We will follow this path until we reach the final target, explaining, as didactically as possible, how it was possible to reach the historic milestone of 100 Mbps with 4G. Let's start with a bit more about Digital Modulations.

Digital Modulation allows information to be sent safely and efficiently, making the most of the technological resources provided by the processors available, which have evolved tremendously in recent years. It was precisely thanks to the production of digital processors with a high degree of integration and the extraordinary increase in processing speed that cellular technology reached the level of excellence it is today. One evolution is the consequence of the other.

Several modulation schemes, already mentioned in previous paragraphs, have been employed. The simplest modulation is PSK (Phase-Shift Keying), as shown in Fig. 4.1, which allows transmission of only one bit 0 or 1. The diagram on the right is called Constellation Diagram and shows us the two possible positions, i.e., the phase of 0° or at 180°.

The modulation most used in the initial systems was QPSK (Quadrature Phase-Shift Keying), which allows, inside the constellation diagram four possible positions or four possible states, or better, with four-phase transitions in 45°, 135°, 225°, or 315°, as shown in Fig. 4.2. In this case, there are four possible positions 11/01/00/10.

The other most common modulations employed are 16-QAM and 64-QAM, as shown in Fig. 4.3. In these cases, we have both phase and amplitude variations.

Table 4.1, shows, for each type of modulation, the number of bits that are generated and transported per transmitted symbol.

© The Author(s), under exclusive license to Springer Nature Switzerland AG 2023
J. L. Frauendorf, É. Almeida de Souza, *The Architectural and Technological Revolution of 5G*, https://doi.org/10.1007/978-3-031-10650-7_4

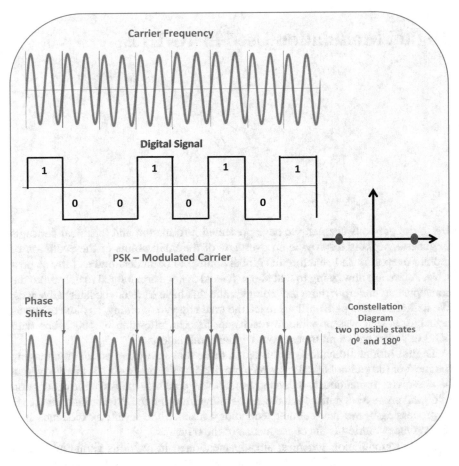

Fig. 4.1 PSK (Phase-Shift Keying) modulation

The symbol corresponds to the shape that the carrier wave presents after modulation and the corresponding states generated. Depending on the number of parameter transitions contained in a period, there will be a corresponding number of information represented by these transitions. The greater the number of possible transitions, the greater the amount of information contained in the symbol. The symbol/second transmission rate is called Baud Rate or Symbol Rate. The data transmission rate/second is called Bit Rate. The correlation between Symbol Rate and Bit Rate is exactly how many bits (n) can be changed or modulated during the 1-second time period of the carrier. Figure 4.4 illustrates this correspondence.

In PSK case, Symbol Rate is equal to Bit Rate. In QPSK, Symbol Rate is half of Bit Rate, and so on. Another way to express this relation is to say that Bit Rate is equivalent to the inverse of Bit Time, that is, it is equivalent to the Transition Period of Bit Time. The Symbol Rate is measured in Symbols/second (Sps).

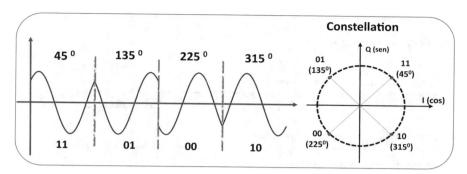

Fig. 4.2 QPSK (Quadrature Phase-Shift Keying) modulation

Table 4.1 Modulation × Bits/symbol

Modulation	Bits/symbol	Number of states
PSK	1	2
QPSK	2	4
16-QAM	4	16
64-QAM	6	64
256-QAM	8	256

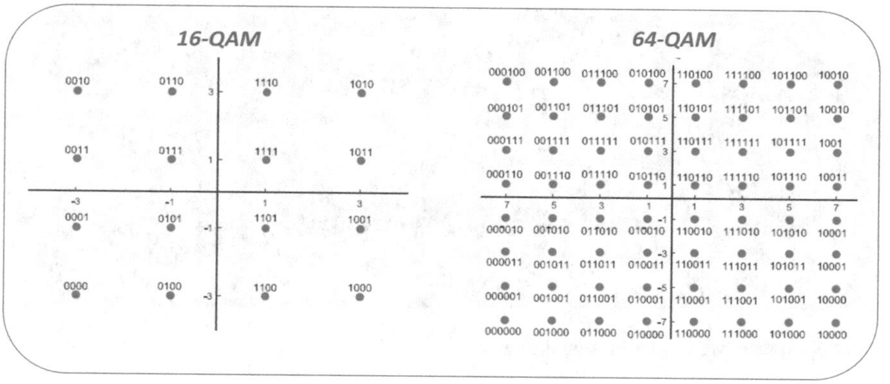

Fig. 4.3 Quadrature Amplitude Modulation (QAM)

We also know that we can divide the radio spectrum into bands and that these bands are suitable for sending information in both directions: from the user device/equipment (UE) to the base station, i.e., uplink, and in the opposite direction (downlink). One can allocate independent *FDD* (Frequency Division Duplex) operation bands for each direction, with the lower band normally used for uplink and the upper band for downlink, or one can allocate part of the transmission time to send information in one direction and part of the total time used to send information in the other direction, *TDD* (Time Division Duplex), as shown in Fig. 4.5.

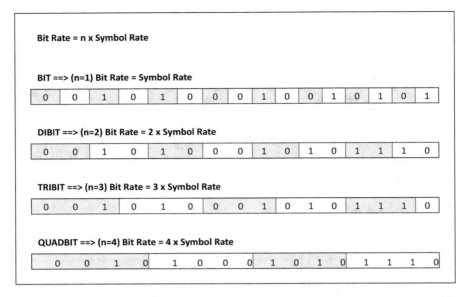

Fig. 4.4 Bit Rate × Symbol Rate correspondence

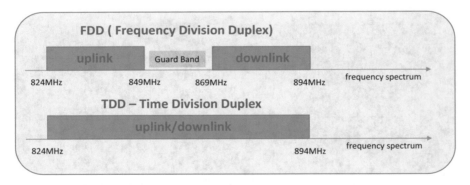

Fig. 4.5 Spectrum division FDD and TDD

Each transmission band is divided into *channels*. Each channel is assigned a carrier frequency, which can operate (excursion) in a small, limited band. Modulation allows one or more parameters of the carrier to be varied so that it carries the information corresponding to the variation or variations.

Each channel, in turn, can be multiplexed. It can be done by sharing the spectrum in subcarriers, which is called frequency spectrum multiplexing, *FDMA* (Frequency Division Multiple Access). The other way to use the spectrum to allow multiplexing is to allocate a portion of the transmission time to a single carrier, called *TDMA* (Time Division Multiple Access). This portion of time allocated to a single carrier is called a *time slot*. Figure 4.6, already depicted, shows the two basic forms of multiplexing and subcarrier configuration.

In the case of dividing the spectrum by multiple frequency channels, orthogonal carriers were used, which have been mentioned before, but which we now explain in detail.

In a conventional spectrum division, each carrier frequency occupies a specific band and maintains a guard band so that it does not cause interference to adjacent channel. In the case of orthogonal carrier division of the spectrum, they are arranged in a compressed fashion, but in a very special way in which, when one carrier is touring the signal at its maximum amplitude, all the others are passing at that very moment at zero value. This allows a much more efficient occupation of the spectrum by saving guard bands. In the time domain, the orthogonality characteristic implies that any two subcarriers differ from each other by an integer number of cycles during a given time interval. This means that they are separated in frequency by a multiple value. Therefore, the first subcarrier is f, the second is $2f$, and so on up to nf, which would be the highest subcarrier used. Figure 4.7 clearly shows how this is possible.

OFDMA (*Orthogonal Frequency-Division Multiple Access*) is a modulation technique based on frequency division multiplexing like FDM. Only in this case, the multiple subcarriers occupying a given spectrum band are orthogonal to each other and modulated independently. Several types of modulation can be used, from QPSK to 256-QAM. The lower the modulation level, the lower the transmission rate. The modulation level is determined by the C/N (Carrier/Noise) ratio. The higher the attenuation of the carrier path, the lower the level received by the receiver and the more difficult it is to detect the transitions carried by the carrier.

The initial concept of OFDMA is to transform a digital signal with high transmission rate into many subcarriers with lower transmission rates and less susceptible to the inherent problems of radio transmission, making the transmission more robust. It would be like dividing the load of a huge truck into smaller loads carried by smaller vehicles that together support the entire load of the larger truck. As the subcarriers are transmitted in parallel, it is possible that each carrier is modulated

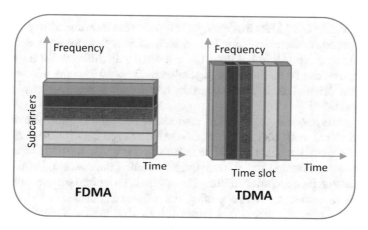

Fig. 4.6 Ways to multiplex a communication channel

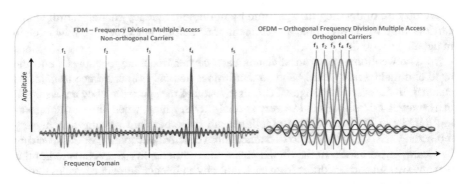

Fig. 4.7 Difference between orthogonal and non-orthogonal carriers

with a lower Symbol Rate than would be required for the original signal. OFDM takes advantage of a concept developed by a French mathematician, Jean-Baptiste Joseph Fourier (1768–1830). Fourier developed a trigonometric series called Fourier Series. The basic concept of this discovery is: *"Any periodic signal can be represented by a sum of harmonically related sinusoidal signals (sine and cosine)."* Each of the signals has specific frequency, phase, and amplitude. With this process, we transform events occurring in the time domain into events occurring in the frequency domain. Figure 4.8 facilitates the understanding of this concept.

With the advance of processors and the development of mathematical algorithms called Fast Fourier Transform (FFT), it was possible to implement this process in specific circuits. This is the case of processors used in both 4G and 5G.

Let's detail how OFDMA modulation is done.

As already mentioned, the OFDMA principle consists in subdividing a signal with high transmission rate into many subcarriers that, by working at a lower transmission rate, are less subject to distortions and interference during propagation. A typical system may have hundreds of orthogonal subcarriers that are equidistant from each other in the radio spectrum at spacings of 10 or 15 kHz. In the case of LTE, the spacing is 15 kHz. Since they are spaced and transmitted in parallel, it is possible that each subcarrier is modulated with a lower Symbol Rate. All subcarriers work with the same Symbol Rate, but can work with different Bit Rates, that is, each subcarrier can be modulated independently. By working with a lower Symbol Rate than the original carrier, the subcarriers become much more robust with respect to propagation.

To make this possible, the OFDMA process begins with the transformation of the modulated signal in QPSK, 16-QAM, 64-QAM, or even 256-QAM, of high transmission rate, in subcarriers that will be transmitted simultaneously in parallel. This process is called Inverse Fourier Transform. Recalling the concept, in this phase of the transmission process, after splitting the original carrier into subcarriers, what we seek is to transform subcarriers operating at multiple frequencies of 15 kHz spacing into a single composite signal that brings together all the individually modulated subcarriers. Figure 4.9 shows in block diagram form how it happens in practice.

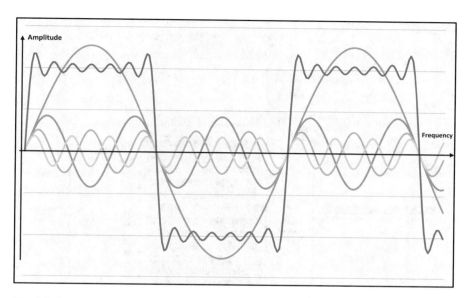

Fig. 4.8 Decomposing a periodic signal using Fourier series

Analyzing the figure, we see that the input signal, a high transmission rate Bit Stream, undergoes the modulation process (QPSK, 16-QAM, 64-QAM, or 256-QAM), whose modulation level is determined by the signal propagation conditions of a particular user, in order to transform bits into modulated symbols. In the sequence, a serial to parallel conversion is performed, which, in the specific case of LTE, separates a set of symbols from blocks of 7 symbols that will modulate one of the 12 subcarriers. This set formed by a block of 7 symbols × 12 subcarriers is called RB (Resource Block). One RB contains, therefore, 84 symbols (Resource Elements) in total.

Let's get a better understanding of what these blocks look like before proceeding. Figure 4.10 shows schematically how the blocks are configured. It is important to remember that these blocks are the carriers of both the user data and the control data needed to communicate between the user unit (UE) and the Base Station efficiently for both uplink and downlink. The LTE design defines that each communication frame should last 10 ms. We have been talking a lot about Packet Networks, Network Package, or even Packet Switched (Packet Switched Data), without having, so far, defined what is a "data packet." Well, a *FRAME* is exactly a set of data that are aggregated/packaged to be transmitted at a certain time interval. This is essential in any data network so that it is possible to transmit respecting the timing of events occurring in a so-called synchronous network, which strictly obeys a protocol or set of rules that define what should occur, and when.

That means, a Frame should last 10 ms and be divided into Subframes of 1 ms, which in turn are composed of 2 Time Slots, as shown in the Fig. 4.10. Each Time

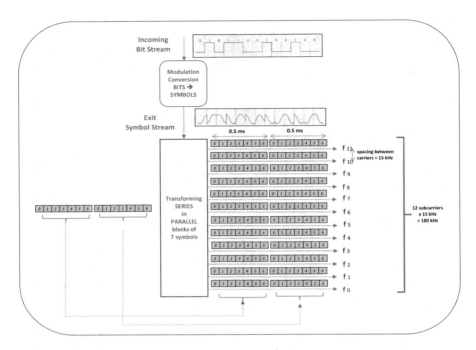

Fig. 4.9 OFDMA LTE concept

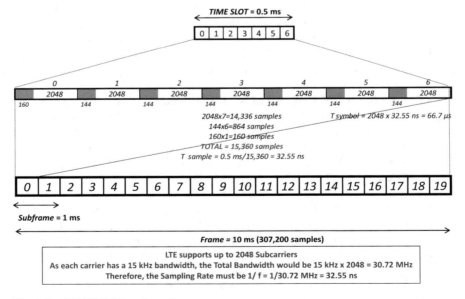

Fig. 4.10 LTE FRAME configuration

Slot contains 6 symbols, and each symbol lasts effectively 66.7 μs (microseconds), or 2048 × 32.55 ns.

How were these numbers defined? We remember that each subcarrier occupies a 15 kHz band, which corresponds to the channel width that a carrier can be modulated without interfering with adjacent orthogonal carriers. In addition, LTE states that it can operate with up to 2048 subcarriers. Therefore, the maximum total range that the spectrum could occupy is 2048 × 15 kHz = 30.72 MHz. Since we know that the cycle time of a wave is the inverse of its frequency, and since 1 MHz = 1,000,000 cycles/second, then the time that lasts 1 cycle of 30.72 MHz is equal to $1/30.72 \times 10^6$ cycles per second, or 32.55 ns (nanoseconds). To this period, we call *Sampling Time (TS)*, that is, how long a sample lasts, and we use it to determine which variation or variations the carrier suffered in this period and then conclude which modulation it suffered. LTE defines, therefore, that each symbol can contain 2048 samplings every 32.55 ns, which totals 66.7 μs, which is the duration time of a part of a Symbol.

Each Symbol, however, has an additional time, that in the case of the first symbol of the Time Slot lasts 5.2 μs (160 × 32.55 ns) and in the others 4.69 μs (144 × 32.55 ns). The fact that the first period is a little bigger than the others is exactly to identify that it is, effectively, the first symbol. This time, which is not considered for the evaluation of the symbol content, was thought to minimize the possible effects of multipaths that a wave may suffer when reflected by some obstacle, in order not to interfere in the evaluation of the modulation of the carrier. A multipath would represent an additional path during signal propagation. The term used to describe the effect is the Delay Spread. Since the speed of light is 300,000 km/s and a wave propagates at exactly that speed, in 4.69 μs a multipath of up to 1.4 km would be tolerated. Multipath means that the signal would arrive late and would interfere with a wave that was received without suffering any reflection. Just as a comment, these times of 5.2 μs, or 4.69 μs are filled with a copy of the sinusoid corresponding to the final stretch of the symbol, just so that the signal level is not zero.

We can finally calculate which Transmission Rates could be achieved in each transmission configuration. Figure 4.11 shows the main parameters needed for this calculation.

We know, therefore, that each Frame (Packet Data) can transmit 168 kS/s (kSymbols/second) and that each Resource Block occupies 180 kHz of spectrum. In addition, the transmission configuration depends on two factors: the spectrum range that is available, which in the case of LTE can be 1.4/3/5/10/15/20 MHz, and the modulation level that the propagation conditions allow, which can be QPSK, 16-QAM or, 64-QAM. Based on these parameters, we can calculate the *Maximum Transmission Rate of LTE* in different combinations of these variables, as shown in Table 4.2. We can verify that LTE's goal was to reach 100 Mbps, which is possible by operating with a spectrum band of 20 MHz and, according to propagation conditions, modulating all carriers in 64-QAM.

Just to conclude the description of the downlink process, it is interesting to show the final stage of transmission, which can be understood by observing Fig. 4.12.

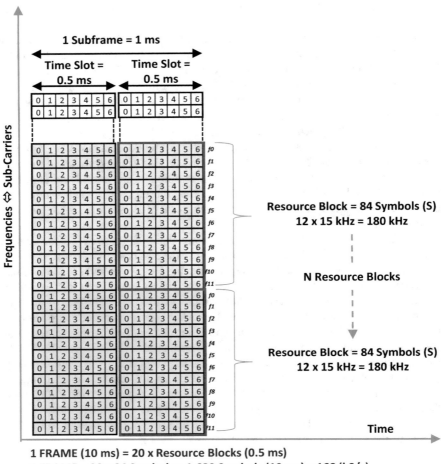

Fig. 4.11 Setting up a Resource Block (RB)

After configuring the Resource Blocks, all subcarriers are converted, using the Inverse Fourier Transform. It is inverse because it converts a signal from the frequency domain to a signal in the time domain. The obtained signal modulates the main carrier, which works exactly on the frequency of the chosen band. It is important to add that Resource Blocks can transport content from several users and for this reason, it is called OFDMA, in which the "A" (access) represents exactly this characteristic. Schematically, we can represent the difference between OFDM and OFDMA observing Fig. 4.13. In OFDM there is a loss of transmission efficiency because each RB is allocated to a single user, even if the traffic volume does not require the use of the entire block.

When receiving the signal, the reverse process occurs.

Table 4.2 LTE maximum transmission rate calculation

Available spectrum width (MHz)	1.4	3	5	10	15	20
Number of resource blocks	6 (6 × 180 = 1.08 MHz)	15 (15 × 180 = 2.7 MHz)	25 (25 × 180 = 4.5 MHz)	50 (50 × 180 = 9.0 MHz)	75 (75 × 180 = 13.5 MHz)	100 (100 × 180 = 18.0 MHz)
Max rate (kbps) QPSK (2 bits/symbol)	2,016	5,040	8,400	16,800	25,200	33,600
Max rate (kbps) 16-QAM (4 bits/symbol)	4,032	10,080	16,800	33,600	50,400	67,200
Max rate (kbps) 64-QAM (6 bits/symbol)	6,048	15,120	25,200	50,400	75,600	100,800

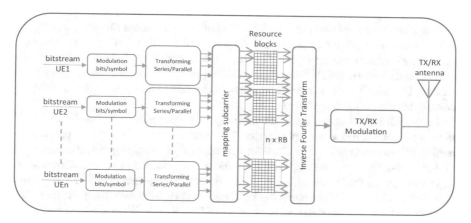

Fig. 4.12 Simplified LTE – OFDMA Diagram

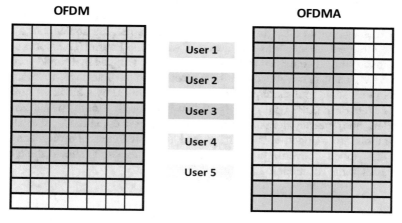

Fig. 4.13 Difference between OFDM and OFDMA

Now that we know OFDMA modulation, used for downstream, let's learn more about SC-FDMA (Single-Carrier Frequency Division Multiple Access), used in upstream.

In SC-FDMA, the symbols are aggregated in order to minimize the amplitude variation of the signals. When amplitude variations are large, as occurs in the transmission of signals generated by base stations, precisely because of the various modulation levels caused by propagation conditions, and the large variation of traffic volume demanded by users, the amplifiers used need to be very linear in a very wide range of levels of the signals being processed. That demands elaborated circuits that make the project more expensive. That's what happens in OFDMA, whose modulation technology is justified because it's more efficient than SC-FDMA and because BS is an equipment that serves many users, but it wouldn't make sense to use OFDMA in equipment used by users. The differences are small, however.

We must remember that the uplink would only be used some subcarriers per user, unlike what happens in the downlink, in which all available subcarriers are used, since they contain messages from all users. Thus, it makes sense in the uplink to use only a set of subcarriers that are responsible for carrying all the information generated by the user. This is achieved by introducing one more stage in the signal processing before being transmitted.

In this additional stage, the symbols are grouped in order to be transported as if they were configured as a single carrier. The way to achieve this is by applying an additional stage of the Fourier Transform, which allows mapping the symbols generated so that they are grouped correctly, and not distributed in several subcarriers, as in the case of OFDMA. SC-FDMA is a form of signal transmission less robust than OFDMA, since in the latter the information is distributed among several subcarriers and not concentrated, as in the case of SC-FDMA. Figure 4.14 shows that. In OFDMA transmission is less susceptible to interference and fluctuations that may occur in selective bands of the spectrum during signal propagation (selective fading).

Figure 4.15 shows how the SC-FDMA carrier is generated. In fact, what is called "*single carrier*" is not a single carrier, but a set of 12 subcarriers that carry the content of messages from a single user. In order to group all the contents of a single user, an intermediate step was introduced in which the Fourier Transform occurs, allowing the content contained in 12 subcarriers to be grouped in a single portion of the time slot. Thus, in this process the data received, in the first stage, are modulated and then forwarded to a serial/parallel converter, followed by a block of the Direct Fourier Transform (DFT), i.e., the signal that is in the time domain will be transformed to the frequency domain. From there on the process is identical to OFDM, the symbols are mapped to subcarriers and then an Inverse Fourier Transform (IDFT) is applied, converting the signal to the time domain again. At the end, the signal goes to the RF module for transmission.

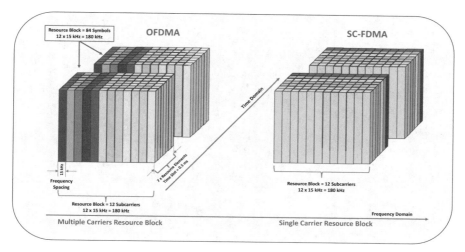

Fig. 4.14 Difference between OFDMA and SC-FDMA

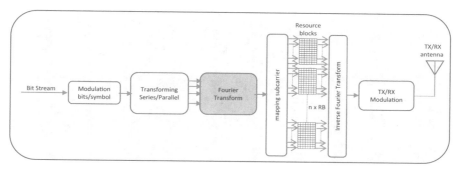

Fig. 4.15 Simplified LTE – SC-FDMA diagram

Chapter 5
Digital Modulation Used in 5G –
Summary: Modulations Used in LTE × 5G

For 20 years, from 1990, when data transmission via cellular systems began, until 2010, when 4G was introduced, it was possible to promote continuous technological advances, which enabled the evolution of transmission rates from 64 kbps to 100 Mbps. Simultaneously, the volume of users served jumped from a few million to over 5 billion. More importantly, a service previously available for a few, became so popular that was accessible to a large part of the world population, possibly resulting the greatest social improvement of that period.

4G was a revolution and opened the doors to a much more ambitious proposal. It broke paradigms and gave new direction to the way of seeing the service to be provided. By adding higher speed, which means quality, and a broad "suite" of mobile applications, LTE has added value to mobile telephony. This has enabled operators to get a better return on the huge investments they have already made and have yet to make and to earn higher profits as revenues from voice telephony have been eroded. But value-added services became more important than transport service. As time went by, mobile operators began to enable profits to those using their transport networks without sharing in the revenues on the products transported.

5G represents a huge evolution. It builds on everything that was revolutionary about 4G and adds new possibilities, making it the structural technology that will underpin the new applications needed for Industry 4.0 and the development of the IoT. Given the huge number of new services and applications, 5G had to become extremely flexible and efficient, being able to serve from extremely simple applications, such as the reading of a temperature sensor that sends data to a platform once or twice a day, to applications that require very high throughput and sensor readings in fractions of a millisecond, such as the control of a mechanical arm. 5G aims to work with extremes, serving the most diverse types of existing technologies and those yet to be created.

To make this possible, the industry had to come together and work intensively to establish common standards and objectives. Several organizations and study groups were engaged in this task, such as ITU-T, GSM, and O-RAN. Once the viability of

© The Author(s), under exclusive license to Springer Nature Switzerland AG 2023
J. L. Frauendorf, É. Almeida de Souza, *The Architectural and Technological Revolution of 5G*, https://doi.org/10.1007/978-3-031-10650-7_5

this new technology was proven in practice, it was up to engineers to explore even more horizons and break paradigms. Three distinct categories of services were created, each one with different objectives and aimed at applications with completely different characteristics, in some cases even with conflicting aspects, as shown in Table 5.1.

- *eMBB* (Enhanced Mobile Broadband) is the service designed to handle large volumes of traffic, with peaks of up to 10–20 Gbps, requiring a minimum throughput of 100 Mbps, supporting mobility that allow speeds of up to 500 km/h.
- *URLLC* (Ultra-Reliable Low-Latency Communications) requires very fast response, <1 ms for the air interface and latency between equipment (E2E) <5 ms. Requires very high connection reliability with uptime of 99.9999%. Provides low and medium transmission rates, which can range from 50 kbps to 10 Mbps. Requires service to vehicles moving at high speed.
- *mMTC* (Massive Machine Type Communications) calls for high device density (2×10^5 or 10^6 /km^2), low data transmission rate, 1–100 kbps, asynchronous mode access and low power consumption by IoT devices to provide up to 10 years lifetime for a single battery.

With the avidity of traffic consumption by users and the gradual exhaustion of available radio frequencies below 3 GHz, it was necessary to seek new spectral bands, wider and at a higher frequency, which also gave a new direction to the morphology of networks. Due to the increase in network densification and higher traffic demand, it was also necessary to change the concept of cellular deployments. Table 5.2 shows the main differences between LTE and 5G.

The higher the frequency used, the more energy must be required to enable its propagation and to overcome a higher degree of attenuation, caused mainly by vegetation, walls, and mirrored glass. Therefore, the 5G network model to be deployed is totally different from the models used until now.

The set of all these factors resulted in the establishment of very well-defined criteria and goals that are summarized in Table 5.3 and that associates Requirements with Applications.

With all these requirements, a lot had to be changed to allow the evolution of LTE to 5G. At first, 5G NR (New Radio) needs to consider a set of more than a dozen parameters whose values must be observed for the proper functioning of the

Table 5.1 New services to be offered by 5G

Requirements to be achieved with the 5G-NR	Services offered
eMBB – Enhanced Mobile Broadband	Gigabyte Internet, 3D video, UHD screen, virtual reality/augmented reality, etc.
URLLC – Ultra Reliable and Low Latency Communication	Self-driving car, mission critical application, industry robot/drone, etc.
mMTC – Massive Machine Type Communication	Smart city, IoT – Internet-of-Things, home automation, eHealth, etc.

Table 5.2 LTE × NG-5G comparison

Evolution LTE x 5G	LTE (3GPP – Rel-8)	NR-5G (3GPP Rel-15)
Operating frequency range	<6 GHz	FR1 – 450 MHz/6 GHz (7.125 GHz) FR2 – 24.25 GHz – 52.6 GHz
Maximum bandwidth – bandwidth (MHz)	20 MHz	FR1: 5, 10, 15, 20, 25, 30, 40, 50, 60, 70, 90, 100 - FR2: 50, 100, 200, 400
Subcarrier spacing (kHz)	15	$2^n \times 15$ for $n = 0,1,2,3,4$ (15,30,60,120,240)
Frequency multiplexing technology	CP-OFDMA (DL) – SC-FDMA (UL)	CP-OFDMA (DL/UL) – DFT-s-OFDMA/ SC-FDMA (UL)
FFT – Fast Fourier Transform	2k (2,048)	2k (2048)/4k (4096)
Maximum modulation levels	Up to 256-QAM(DL)/64-QAM (UL)	Up to 256-QAM (DL/UL)
Maximum number of subcarriers	1200	3276
Subframe duration (subframe length) (ms)	1 (0.5)	1
Air interface latency (ms)	10	1
Slot size	7 symbols in 0.5 ms	14 symbols (duration depends on subcarrier spacing =1/0.5/0.25/0.125/0.0625 ms)/ mini-SLOTS 2, 4 and 7 symbols
Antenna support beamforming	No	Yes
Antenna Support massive MIMO	Up to 8T/8R	4T/4R essential/can reach up to 64T/64R
Duplexing format	FDD/TDD – static	FDD/TDD – static/dynamic

connected equipment. To handle all this diversity, NR uses several types of subcarriers spacing to distribute resources according to service needs.

This technique of considering multiple "subcarrier spacing types" is called *Numerology*. Numerology allows the 5G Base Radio Station (known as gNB) to allocate radio resources in a more flexible way.

The numerology obeys the following equation: (2^n), where "n" can assume values from 0 to 5. It is important to maintain this ratio of multiples of the value of 15 kHz, which is the basic frequency used by LTE, because orthogonality between the subcarriers is maintained. Table 5.4 contains the main values that are imputed to the parameters as a result of the values assumed by numerology. Note that the value of "n equal to 5" is shown, although this spacing option is not being used at this time and, therefore, we will make a few comments about it.

This table is extensive, but extremely useful due to the plurality that the variables can assume as a result of the numerology chosen. Based on this table, we will make

Table 5.3 *5G* requirements

Parameter	Requirements	Application
Peak-data-rate(maximum peak value)	DL – 20Gbps/UL – 10Gbps	eMBB
Spectral efficiency	DL – 30 bits/Hertz/UL – 15 bits/Hertz	eMBB
Latency	C-Plane – 10 ms/U-Plane – 0.5 ms	URLLC
User speed	DL – 100 Mbps/UL – 50 Mbps	eMBB
Capacity/traffic density	10 Mbits/m²	mMTC
Connection density	1 million devices/km²	mMTC
Battery power efficiency	90% reduction of energy consumption	mMTC
Transmission reliability	1 Packet lost/100 million Packets transmitted	URLLC
Mobility	500 km/h	eMBB
Disconnection time during mobility	ZERO	URLLC
Maximum spectrum range (bandwidth)	Up to 1 GHz	eMBB
Coverage	164 dB	mMTC
Battery life-time	10 years	mMTC

the following considerations, carefully analyzing these parameters and how they are evolving from LTE to 5G.

Let's start by evaluating carrier spacing. As the 5G coverage range is very wide, it has been divided into two bands, FR1 and FR2, with FR1 being the bands below 6 GHz and FR2 being the radio frequency bands above 6 GHz, as shown in detail in Table 5.5, just a slice of Table 5.4 in which we highlight the main similarities and differences between LTE and 5G. One could say that $n = 0$ is equal to LTE.

Increase in the Number of Subcarriers

In LTE, the carrier spacing is single, 15 kHz, and the maximum number of subcarriers is 2048 (Fast Fourier Transform of 2k). With this, the maximum range of spectrum that could support this system would be 30.72 MHz (2048 × 15 kHz). However, the maximum range used in LTE is 20 MHz, which limits the number of subcarriers to 1200 (15 kHz × 1200 = 18 MHz). To achieve wider spectrum utilization, the number of subcarriers must be increased from 2048 to 4096.

Increase in Subcarrier Spacing

Just increasing the number of subcarriers was not enough to increase the spectrum occupancy, but it was also necessary to increase the spacing between subcarriers from 15 kHz to values of 30, 60, 120, and even 240 kHz, expected to reach 480 kHz.

Table 5.4 Numerology and main parameters of 5G

Numerology n =	0	1	2	3	4	5	LTE
(2^n) =	1	2	4	8	16	32	NA
FR1 ==> 450 MHz – 6 GHz frequency range	Yes	Yes	Yes	No	No	No	NA
FR2 ==> 24.25 GHz – 52.6 GHz frequency range	No	No	No	Yes	Yes	???	NA
Space between subcarriers (kHz)	15	30	60	120	240	480	15
RB – resource block size – 12 subcarriers/RB (kHz)	180	360	720	1440	2880	5760	180
Maximum bandwidth (MHz) for 2048 subcarriers (FFT 2k)	30.72	61.44	122.88	245.76	491.52	983.04	30.72
Maximum bandwidth (MHz) for 4096 subcarriers (FFT 4k)	61.44	122.88	245.76	491.52	983.04	1966.08	NA
Ts – sampling time – FFT 2k (ns)	32.552	16.276	8.138	4.069	2.035	1.017	32.552
Tc – sampling time – FFT 4k (ns)	16.276	8.138	4.069	2.035	1.017	0.509	NA
Frame duration (ms)	10	10	10	10	10	10	10
Number of slots/frame	10	20	40	80	160	320	10
Subframe duration (ms)	1	1	1	1	1	1	1
Number of slots/frame	10	20	40	80	160	320	10
Number of slots/subframe	1	2	4	8	16	32	1
Slot duration (ms)	1	0.5	0.25	0.125	0.0625	0.03125	1
Symbol duration = 1/slot length(μs)	66.67	33.33	16.67	8.33	4.17	2.08	66.67
Slot size = number of symbols/slot	14	14	14	14	14	14	14
Number of symbols/physical resource block (PRB = 14 symbols/resource elements × 12 subcarriers)	168	168	168	168	168	168	168
Number PRB/subframe	1	2	4	8	16	32	1
Number of symbols/subframe	168	336	672	1344	2688	5376	168
Number of symbols/frame	1680	3360	6720	13,440	26,880	53,760	1680
CP cyclic prefix – long symbol (μs)	5.2	2.86	1.69	1.11	0.81	0.41	5.2
Number of samples long CP FFT2k	160	176	208	273	398	398	160
Number of samples long CP FFT4k	319	351	415	546	796	796	NA
Maximum multipath – long CP (m)	1560	858	507	333	243	122	1560
CP cyclic prefix – Normal symbol (μs)	4.69	2.34	1.17	0.59	0.29	0.15	4.69
Number of samples normal CP FFT2k	144	144	144	145	143	143	144
Number of samples normal CP FFT4k	288	288	288	290	285	285	NA
Maximum multipath – long CP (m)	1407	702	351	177	87	44	1407
Long CP number/subframe	2	2	2	2	2	2	2

(continued)

Table 5.4   (continued)

Numerology n =	0	1	2	3	4	5	LTE
(2^n) =	1	2	4	8	16	32	NA
Number of normal CP/subframe	12	12	12	12	12	12	12
CP overhead (%)	6.67	6.75	6.95	7.38	8.04	8.04	6.67
OFDM symbol + long CP (μs)	71.87	36.19	18.36	9.44	4.98	2.49	71.87
OFDM symbol + Normal CP (μs)	71.36	35.67	17.84	8.92	4.46	2.23	71.36
Minimum scheduling interval = Total slot duration (14 symbols + 2 long CPs + 12 normal CPs) (ms)	1	0.500	0.251	0.126	0.063	0.032	1

Table 5.5   Key differences between LTE and 5G

Numerology n=	0	1	2	3	4	5	LTE
(2^n) =	1	2	4	8	16	32	NA
FR1 ==> Frequency range 450MHz - 6GHz	Yes	Yes	Yes	No	No	No	NA
FR2==> Frequency range 24.25GHz - 52.6GHz	No	No	No	Yes	Yes	???	NA
Spacing between subcarriers (Slot Length) (kHz)	15	30	60	120	240	480	15
Resource block size – 12 subcarriers/RB (kHz)	180	360	720	1,440	2,880	5,760	180
Maximum bandwidth (MHz) for2048 subcarriers (FFT 2k)	30.72	61.44	122.88	245.76	491.52	983.04	30.72
Maximum bandwidth (MHz) for4096 subcarriers (FFT 4k)	61.44	122.88	245.76	491.52	983.04	1,966.08	NA
Ts - sampling time FFT 2k (ns)	32.552	16.276	8.138	4.069	2.035	1.017	32.552
Tc - sampling time FFT 4k (ns)	16.276	8.138	4.069	2.035	1.017	0.509	NA

By doing so, it is possible to reach the mark of almost 1 GHz of spectrum width for the highest expected operating frequency of 5G (240 kHz × 4096 = 983.04 MHz). The maximum number of subcarriers that can be allocated for 5G NR is 3276. It would even be possible to get a larger bandwidth of 240 kHz × 3276 = 786 MHz. However, if we consider the 400 MHz band as the maximum width, we have 120 kHz × 3276 = 393 MHz. It is important to observe that, for the band <6 GHz, the maximum bandwidth is 200 MHz (60 kHz × 4096 = 245.76 MHz).

Figure 5.1 graphically shows the expansion of subcarrier spacing that evolved from 15 kHz in LTE to 30, 60, 120, and 240 kHz in the 14 symbols per Subframe configuration, and 60 kHz for the 12 symbols per Subframe configuration. This special configuration will be explained later. Figure 5.1 also shows that numerology determines the reduction of the TIME SLOT to increase the number of symbols that can be processed within a Subframe.

It is important to point out that both the spacing of the subcarriers and the decrease of the TIME SLOT are due solely and exclusively to the fact that they follow a rule imposed by numerology. The increase in spacing between subcarriers is fundamental for reasons that will be explained next.

As 5G must operate in higher frequencies, millimeter wave (mW), it is necessary to consider that the circuits that generate the high frequency carrier wave are more critical and tend to vary the phase of the generated wave. This change in phase could cause interference between symbols if the spacing between carriers was maintained. By increasing the spacing, this problem is minimized and allows greater tolerance

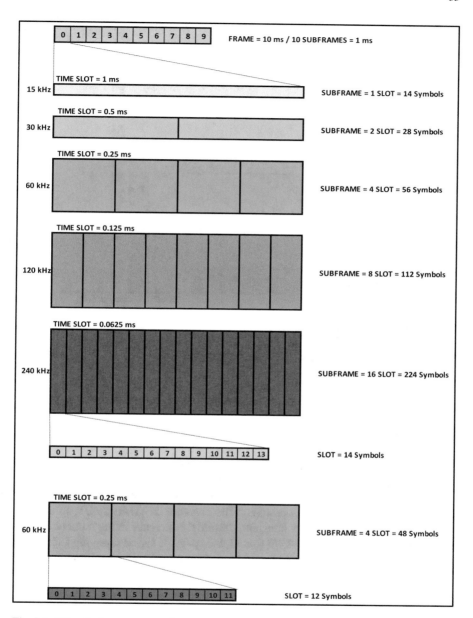

Fig. 5.1 5G radio frame × numerology

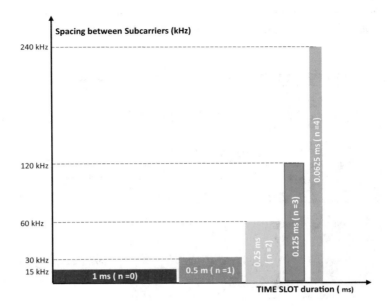

Fig. 5.2   5G numerology (2^n, $n = 0,1,2,3,4$)

in the design of local oscillators. Figure 5.2 shows that, by increasing the spacing between carriers, the symbol duration decreases.

Sampling Rate × Sampling Time (Ts/Tc)

It is important to note the huge difference between the sampling times, which in LTE is called *Ts*, and in 5G, *Tc*. While in LTE the *Ts* value is 32.552 ns, in 5G the *Tc* can reach 1.017 ns, that is, 32 times smaller. This can only be achieved by the huge advance of signal processors. With the decrease of TIME SLOT, so that the number of samples can be kept constant, despite the increase in the number of sub-carriers from 2048 for LTE to 4096 for 5G, the speed of signal sampling (Sampling Rate) must increase and the spacing between two consecutive samples (Sampling Time) must decrease.

Setting the SLOT and Its TIME SLOT

Let's evaluate how all this results in increasing the capacity of the system. First, let's remember what was explained in 4G regarding a Frame, that is, the size of the Packet Data. 5G should maintain compatibility with LTE so that the two technologies can coexist. For this reason, the main parameters of LTE should be preserved, i.e., the Frame duration remains 10 ms, and a Subframe should last 1 ms. According to numerology, the decrease of the duration time of a symbol varies and, with this, the number of Resource Elements (which is equal to a symbol), and the number of SLOTS, which can be transmitted in the available time of a Subframe, or of a Frame, increases the capacity exponentially (2^n), reaching 2, 4, 8, 16, or even 32 times the LTE capacity.

In the case of the 60 kHz spacing, 5G has an optional configuration where the number of symbols per SLOT is not 14 (Normal CP), but 12 (Extended CP), and is considered a special case, which is worth mentioning as it was shown in Fig. 5.1.

Figure 5.3 shows in detail the composition of a SLOT of 14 symbols for each subcarrier spacing, considering the number of samples intended for CPs (Cyclic Prefix) Long and Normal, and the number of samples referring to each symbol (2048 or 4096), with the respective Sampling Time (Ts or Tc). Ts and Tc, as already explained, correspond exactly to the duration of a sampling for each frequency spacing. Adding the total number of samplings and multiplying the total by Ts or Tc, for each case, we obtain the duration time of a SLOT the SLOT Duration. The SLOT Duration is not influenced by the number of subcarriers used (2048 or 4096) for the case of 15 kHz spacing.

We must now remember the existence of *Cyclic Prefix*, symbol stretch that is added to the symbol duration to minimize the influence of multipath and interference between symbols. We must also remember that there are two types of CPs, a Long CP is inserted at the beginning of SLOT and another between the seventh and eighth symbols (to maintain compatibility with LTE), while the other symbols suffer only the addition of the Normal CP (Short CP). These times are neglected in the reception and added only so that there is no transition to zero of the subcarrier value in the separation of the duration times between symbols. The Long CP is important to mark the beginning and the middle of the SLOT, highlighting the start duration of the other symbols.

We can observe in Table 5.6 something very important. As the subcarrier spacing increases, the duration time of both the Long CP and Normal CP varies, and with that, the maximum distance that is supported due to multipath is also changed. This calculation is done based on the propagation speed of the radio wave, which is the same as the speed of light, as already shown for the LTE case. The multipath is more critical when dealing with lower frequencies <6 GHz, with longer wavelengths, and these frequencies are used in cells with larger radii, since they suffer less attenuation. In the case of higher frequencies, used in smaller cells, where radio frequencies are more attenuated, multipath interference loses its influence because signals are more directional.

Subcarrier Frequency	Tc/Ts Sampling Time (ns)	SLOT Duration (ms)	Samples	Long CP	Symbol 0	Short CP	Symbol 1	Short CP	Symbol 2	Short CP	Symbol 3	Short CP	Symbol 4	Short CP	Symbol 5	Short CP	Symbol 6	Long CP	Symbol 7	Short CP	Symbol 8	Short CP	Symbol 9	Short CP	Symbol 10	Short CP	Symbol 11	Short CP	Symbol 12	Short CP	Symbol 13
15 kHz FFT2k	Ts = 32.552	1	Number of Samples	320	2,048	288	2,048	288	2,048	288	2,048	288	2,048	288	2,048	288	2,048	320	2,048	288	2,048	288	2,048	288	2,048	288	2,048	288	2,048	288	2,048
15kHz FFT4k	Tc = 16.276	1	Number of Samples	320	4,096	288	4,096	288	4,096	288	4,096	288	4,096	288	4,096	288	4,096	320	4,096	288	4,096	288	4,096	288	4,096	288	4,096	288	4,096	288	4,096
30kHz FFT4k	Tc = 8.138	0.5	Number of Samples	320	4,096	288	4,096	288	4,096	288	4,096	288	4,096	288	4,096	288	4,096	320	4,096	288	4,096	288	4,096	288	4,096	288	4,096	288	4,096	288	4,096
60kHz FFT4k	Tc = 4.069	0.25	Number of Samples	320	4,096	288	4,096	288	4,096	288	4,096	288	4,096	288	4,096	288	4,096	320	4,096	288	4,096	288	4,096	288	4,096	288	4,096	288	4,096	288	4,096
120kHz FFT4k	Tc = 2.035	0.125	Number of Samples	320	4,096	288	4,096	288	4,096	288	4,096	288	4,096	288	4,096	288	4,096	320	4,096	288	4,096	288	4,096	288	4,096	288	4,096	288	4,096	288	4,096
240kHz FFT4k	Tc = 1.017	0.0625	Number of Samples	320	4,096	288	4,096	288	4,096	288	4,096	288	4,096	288	4,096	288	4,096	320	4,096	288	4,096	288	4,096	288	4,096	288	4,096	288	4,096	288	4,096

Fig. 5.3 Formation of a SLOT composed of 14 Symbols × SLOT duration time

Table 5.6 Maximum multipath according to 5G numerology

Numerology n=	0	1	2	3	4	LTE
$(2^n)=$	1	2	4	8	16	NA
FR1 ==> frequency range 450 MHz - 6 GHz	Yes	Yes	Yes	No	No	NA
FR2 ==> frequency range 24.25 GHz - 52.6 GHz	No	No	No	Yes	Yes	NA
CP - Cyclic Prefix long symbol (µs)	5.2	2.86	1.69	1.11	0.81	5.2
Number of samples long CP FFT2k	160	176	208	273	398	160
Number of samples long CP FFT4k	319	351	415	546	796	NA
Maximum multipath - long CP (m)	1,560	858	507	333	243	1,560
CP - Cyclic Prefix normal symbol (µs)	4.69	2.34	1.17	0.59	0.29	4.69
Number of samples normal CP FFT2k	144	144	144	145	143	144
Number of samples normal CP FFT4k	288	288	288	290	285	NA
Maximum multipath - normal CP (m)	1,407	702	351	177	87	1,407
Long CP number/subframe	2	2	2	2	2	2
Number of normal CP/subframe	12	12	12	12	12	12
CP overhead (%)	6.67	6.75	6.95	7.38	8.04	6.67
Total duration = OFDM symbol + long CP (µs)	71.87	36.19	18.36	9.44	4.98	71.87
Total duration = OFDM symbol + normal CP (µs)	71.36	35.67	17.84	8.92	4.46	71.36
Minimum scheduling interval = Total duration of slot (14 Symbols + 2 long CPs + 12 short CPs) (ms)	1	0.500	0.251	0.126	0.063	1

System Capacity According to Subcarrier Spacing

The increase of the spacing between subcarriers and the respective decrease of the SLOT duration brings as consequence the increase of the number of SLOTS inside a Subframe that lasts 1 ms. What is equivalent to say that the duration time of a SLOT decreases, as already seen above, which is extremely desirable for the latency decrease. At this moment the *PRB* (Physical Resource Block) appears, constituted by a block of 14 Symbols (Time Axis) × 12 Subcarriers (Frequency Axis), as shown in Fig. 5.4.

See that in the case of 15 kHz spacing, there is a single PRB within a Subframe that lasts 1 ms, but this number increases to 16 when using 240 kHz spacing. In this case, the number of symbols within a PRB jumps from 168 to 2688, i.e., 16 times, as shown in Fig. 5.4 and Table 5.7.

With this, it is possible to calculate the Transmission Rate of a PRB, which can reach up to 2.688 M Symbols/s versus 0.168 M Symbols/s in the case of LTE. Therefore, the Transmission Rate can be calculated according to the available spectrum range and subcarrier spacing, as shown in the following tables.

Calculation of the Maximum Transmission Rate According to Numerology and Available Spectral Range

We have almost all the elements to make this calculation, but we are still missing one aspect not yet addressed, the *MGB* (Minimum Guard Band). This Guard Band is necessary and specified by standard to avoid interference between systems operating in the same region of the spectrum, as explained earlier. The MGB is specified for both FR1 and FR2 and depends on the bandwidth available for operation. Knowing the width of a Resource Block, which is equivalent to a set of 12

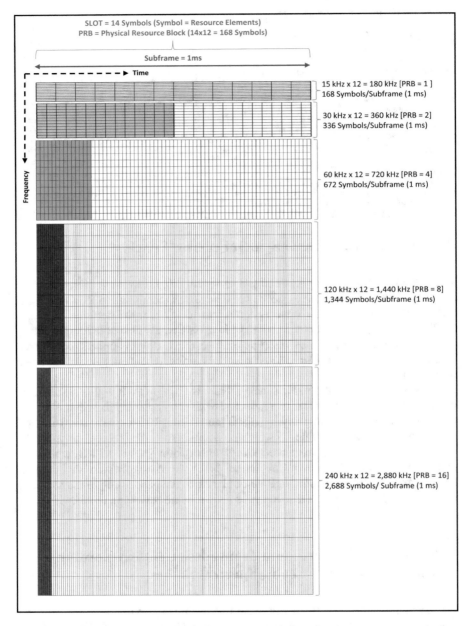

Fig. 5.4 *Subframe* structure × *PRB (Physical Resource Block)*

subcarriers, and the various MGB defined by standards, we have as a result the number of RB available for each band, as shown in Tables 5.8 and 5.9, prepared for both FR1 and FR2.

Table 5.7 **Transmission rate** of a PRB – *Physical Resource Block*

Numerology n=	0	1	2	3	4	LTE
$(2^n)=$	1	2	4	8	16	NA
FR1 ==> Frequency range 450 MHz - 6 GHz	Yes	Yes	Yes	No	No	NA
FR2 ==> Frequency range 24.25 GHz - 52.6 GHz	No	No	No	Yes	Yes	NA
Duration of frame (ms)	10	10	10	10	10	10
Number of slots/frame	10	20	40	80	160	10
Duration of subframe (ms)	1	1	1	1	1	1
Number of slots/subframe	1	2	4	8	16	1
Duration of one slot(ms)	1	0.5	0.25	0.125	0.0625	1
Symbol duration = 1/slot length (μs)	66.67	33.33	16.67	8.33	4.17	66.67
Slot size number of symbols/slot	14	14	14	14	14	14
Number of symbols /physical resource block (PRB = 14 symbols/resource elements x 12 subcarriers)	168	168	168	168	168	168
PRB number/subframe	1	2	4	8	16	1
Number of symbols/subframe	168	336	672	1,344	2,688	168
Number of symbols/frame	1,680	3,360	6,720	13,440	26,880	1,680
PRB transmission rate (M symbols/s)	0.168	0.336	0.672	1.344	2.688	0.168

Table 5.8 Maximum number of Resource Blocks (#RB) according to the Minimum Guard Band (MGB) and according to the Numerology and Spectrum Range available (FR1 < 6 GHz)

Numerology (2^n)	SCS subcarrier spacing (kHz)	MHz ==>	5	10	15	20	25	30	40	50	60	70	80	90	100
n=0	15	MGB (kHz)	242.5	312.5	382.5	452.5	522.5	592.5	552.5	692.5	NA	NA	NA	NA	NA
		RB width (kHz)	180	180	180	180	180	180	180	180	180	180	180	180	180
		# RB	25	52	79	106	133	160	216	270	NA	NA	NA	NA	NA
n=1	30	MGB (kHz)	505	665	645	805	785	945	905	1045	825	965	925	885	845
		RB width (kHz)	360	360	360	360	360	360	360	360	360	360	360	360	360
		# RB	11	24	38	51	65	78	106	133	162	189	217	245	273
n=2	60	MGB (kHz)	NA	1010	990	1330	1310	1290	1610	1570	1530	1490	1450	1410	1370
		RB width (kHz)	720	720	720	720	720	720	720	720	720	720	720	720	720
		# RB	NA	11	18	24	31	38	51	65	79	93	107	121	135

Table 5.9 Maximum number of Resource Blocks (#RB) according to the Minimum Guard Band (MGB) and according to the Numerology and Spectrum Range available (FR2 > 6 GHz)

Numerology (2^n)	SCS subcarrier spacing (kHz)	MHz ==>	50	100	200	400
n=2	60	MGB (kHz)	1210	2450	4930	NA
		RB width (kHz)	720	720	720	720
		# RB	66	132	264	NA
n=3	120	MGB (kHz)	1900	2420	4900	9860
		RB width (kHz)	1440	1440	1440	1440
		# RB	32	66	132	264
n=4	240	MGB (kHz)	NA	3800	7720	15560
		RB width (kHz)	2880	2880	2880	2880
		# RB	NA	43	86	128

Based on this data, it is possible to calculate the minimum and maximum number of RB for each numerology, as shown in Table 5.10.

Table 5.10 Minimum and Maximum amplitude of the transmission band (BW), considering the Guard Bands, and Minimum and Maximum Number of RB

Numerology (2^n)	SCS subcarrier spacing (kHz)	Smallest number of RB	Smallest channel BW (MHz)	Largest number of RB	Largest channel BW(MHz)
$n = 0$	15	25	4.5	270	48.6
$n = 1$	30	24	8.6	273	98.3
$n = 2$	60	24	17.3	264	190.1
$n = 3$	120	32	46.1	264	380.2

Table 5.11 Maximum Transmission Rate for the greatest number of permissible RBs

Numerology (2^n)	SCS subcarrier spacing (kHz)	Number of symbols /resource block	Largest number of RB	Maximum number of symbols (1ms)	QPSK 2 bits / symbol (Mbps)	16-QAM 4 bits / symbol (Mbps)	64-QAM 6 bits / symbol (Mbps)	256-QAM 8 bits / symbol (Mbps)
LTE	15	168	100	16,800	34	67	101	AT
$n=0$	15	168	270	45,360	91	181	272	363
$n=1$	30	336	273	91,728	183	367	550	734
$n=2$	60 (FR1)	672	135	90,720	181	363	544	726
$n=2$	60 (FR2)	672	264	177,408	355	710	1,064	1,419
$n=3$	120	1344	264	354,816	710	1,419	2,129	2,839

Finally, it is possible to perform the calculation of the Maximum Rates for the various numerologies occupying the widest available range for both FR1 and FR2, as shown in Table 5.11.

We can observe the evolution of LTE occupying a band of 20 MHz and using the highest modulation available, 64-QAM, which allows reaching the speed of 100 Mbps. However, the 5G occupying a spectrum band of 400 MHz can reach, with a subcarrier spacing of 120 kHz and maximum modulation level of 256-QAM, the speed of 2.8 Gbps. Based on these calculations, it is possible to know the Spectral Efficiency achieved in these conditions, as shown in Table 5.12.

There are some peculiarities that differentiate LTE from 5G, or rather, make 5G more flexible than LTE. Although the use of both FDD and TDD is planned, due to flexibility and gains in terms of spectrum utilization, it makes sense to use TDD almost exclusively. It is always important to remember that the use of TDD allows to evaluate, almost continuously, the propagation conditions, since the same portion of the spectrum is used for both directions of transmission.

The modulation system used in LTE is OFDMA for DL and SC-FDMA for UL. In 5G, OFDMA is called *CP-OFDMA* (CP = Cyclic Prefix) and can be used both in DL and UL. SC-FDMA can also be used for UL, which in 5G is called *DFT-S-FDMA*, since it has one more step than CP-FDMA, which is *DFT-S* (Direct Fourier Transform-Spread).

5G introduces a new concept in the system, the so-called BWP (Band Width Parts), i.e., partition of the available band in several sections. Each segment can work with a different numerology to better meet the specific needs of simultaneous

Table 5.12 Spectral efficiency for maximum transmission rate/spectrum bandwidth

Numerology (2^n)	SCS subcarrier spacing (kHz)	Spectrum bandwidth (MHz)	QPSK 2 bits/ Symbol (b/ Hz)	16-QAM 4 bits/ Symbol (b/ Hz)	64-QAM 6 bits/ Symbol (b/ Hz)	256-QAM 8 bits/ Symbol (b/Hz)
LTE	15	20	1.68	3.36	5.04	NA
$n = 0$	15	50	1.81	3.63	5.44	7.26
$n = 1$	30	100	1.83	3.67	5.50	7.34
$n = 2$	60 (FR1)	100	1.81	3.63	5.44	7.26
$n = 2$	61 (FR2)	200	1.77	3.55	5.32	7.10
$n = 3$	120	400	1.77	3.55	5.32	7.10

traffic. This concept comes from the need to slice the services provided, the so-called Network Slicing, i.e., the slicing of the network to meet services with different requirements, as already discussed. Thus, the system can handle different services harmoniously. As those that require high data volumes and require a smaller spacing between subcarriers, regardless of latency, while it can handle other data streams generated by devices of lower complexity that require lower data volumes and greater spacing between carriers, but for which latency may be something relevant. To minimize interference between the various portions of the spectrum, allocated to different services, are used Blank Slots, i.e., slots that are not programmed in the transmission, thus creating a separation between the various services provided.

A single user device (UE) can accommodate up to four distinct portions of spectrum, each of which can be allocated to different services to meet both DL and UL demand, with each portion activated at different times. This allows for flexible service provisioning, meeting Network Slicing requirements.

This flexibility allows, for example, the system to be configured to optimize the daytime traffic volume demand (eMBB) in order to meet the demand of a larger number of IoT devices (mMTC) at nighttime. In the case of eMBB, the demand tends to be higher in the DL direction than in the UL. As for mMTC, the demand may be exactly the opposite.

5G allows each symbol in a frame to be flexibly allocated, both in the DL and UL, depending on the need. For this, a mechanism called *SFI* (Slot Format Indicator) is used, in a statistical way, being able to vary its configuration dynamically to respond instantaneously to the network needs. Up to now there are 55 possible formats for symbol allocation within a SLOT.

This is how 5G can meet the requirements of the New RAN (New Radio Access Network). These properties make the Physical Layer, as we will see in a later chapter, flexible and scalable, i.e., it can adapt to meet the various services that make up Network Slicing, allowing users to contract exactly the service that meets their needs. This differentiates 5G from all previous systems.

In addition, to meet the requirement of reducing latency time, there are several ways to configure a SLOT, allocating less than 14 symbols to the uplink or downlink allowing greater flexibility in transmission, depending on the instantaneous need.

Table 5.13   SFI (Slot Format Indicator) D = Downlink / U = Uplink / F = Flexible

Format	Symbol number in a slot													
	0	1	2	3	4	5	6	7	8	9	10	11	12	13
0	D	D	D	D	D	D	D	D	D	D	D	D	D	D
1	U	U	U	U	U	U	U	U	U	U	U	U	U	U
2	F	F	F	F	F	F	F	F	F	F	F	F	F	F
3	D	D	D	D	D	D	D	D	D	D	D	D	D	F
4	D	D	D	D	D	D	D	D	D	D	D	D	F	F
5	D	D	D	D	D	D	D	D	D	D	D	F	F	F
6	D	D	D	D	D	D	D	D	D	D	F	F	F	F
7	D	D	D	D	D	D	D	D	D	F	F	F	F	F
8	F	F	F	F	F	F	F	F	F	F	F	F	F	U
9	F	F	F	F	F	F	F	F	F	F	F	F	U	U
10	F	U	U	U	U	U	U	U	U	U	U	U	U	U
11	F	F	U	U	U	U	U	U	U	U	U	U	U	U
12	F	F	F	U	U	U	U	U	U	U	U	U	U	U
13	F	F	F	F	U	U	U	U	U	U	U	U	U	U
14	F	F	F	F	F	U	U	U	U	U	U	U	U	U
15	F	F	F	F	F	F	U	U	U	U	U	U	U	U
16	D	F	F	F	F	F	F	F	F	F	F	F	F	F
17	D	D	F	F	F	F	F	F	F	F	F	F	F	F
18	D	D	D	F	F	F	F	F	F	F	F	F	F	F
19	D	F	F	F	F	F	F	F	F	F	F	F	F	U
20	D	D	F	F	F	F	F	F	F	F	F	F	F	U
21	D	D	D	F	F	F	F	F	F	F	F	F	F	U
22	D	F	F	F	F	F	F	F	F	F	F	F	U	U
23	D	D	F	F	F	F	F	F	F	F	F	F	U	U
24	D	D	D	F	F	F	F	F	F	F	F	F	U	U
25	D	F	F	F	F	F	F	F	F	F	F	U	U	U
26	D	D	F	F	F	F	F	F	F	F	F	U	U	U
27	D	D	D	F	F	F	F	F	F	F	F	U	U	U
28	D	D	D	D	D	D	D	D	D	D	D	D	F	U
29	D	D	D	D	D	D	D	D	D	D	D	F	F	U
30	D	D	D	D	D	D	D	D	D	D	F	F	F	U
31	D	D	D	D	D	D	D	D	D	D	D	F	U	U
32	D	D	D	D	D	D	D	D	D	D	F	F	U	U
33	D	D	D	D	D	D	D	D	D	F	F	F	U	U
34	D	F	U	U	U	U	U	U	U	U	U	U	U	U
35	D	D	F	U	U	U	U	U	U	U	U	U	U	U
36	D	D	D	F	U	U	U	U	U	U	U	U	U	U
37	D	F	F	U	U	U	U	U	U	U	U	U	U	U
38	D	D	F	F	U	U	U	U	U	U	U	U	U	U
39	D	D	D	F	F	U	U	U	U	U	U	U	U	U
40	D	F	F	F	U	U	U	U	U	U	U	U	U	U
41	D	D	F	F	F	U	U	U	U	U	U	U	U	U
42	D	D	D	F	F	F	U	U	U	U	U	U	U	U
43	D	D	D	D	D	D	D	D	D	F	F	F	F	U
44	D	D	D	D	D	D	F	F	F	F	F	F	U	U
45	D	D	D	D	D	D	F	F	U	U	U	U	U	U
46	D	D	D	D	D	F	U	D	D	D	D	D	F	U
47	D	D	F	U	U	U	U	D	D	F	U	U	U	U
48	D	F	U	U	U	U	U	D	F	U	U	U	U	U
49	D	D	D	D	F	F	U	D	D	D	D	F	F	U
50	D	D	D	F	F	U	U	D	D	D	F	F	U	U
51	D	F	F	U	U	U	U	D	F	F	U	U	U	U
52	D	F	F	F	F	F	U	D	F	F	F	F	F	U
53	D	D	F	F	F	F	U	D	D	F	F	F	F	U
54	F	F	F	F	F	F	F	D	D	D	D	D	D	D
55	D	D	F	F	F	U	U	U	D	D	D	D	D	D
56 – 254	Reserved													
255	UE determines the slot format for the slot based on *tdd-UL-DL-ConfigurationCommon*, or *tdd-UL-DL-ConfigurationDedicated* and, if any, on detected DCI formats													

Source: 3GPP TS 38.213 version 15.7.0 Release 15

This can occur dynamically. Table 5.13 shows the various configurations of the *Slot Format Indicator (SFI)* that are already provided, allocating more, or less, symbols to the DL or UL.

In LTE, a Subframe, which coincides with a SLOT in 5G NR, is configured for either DL or UL, i.e., all symbols in the SLOT/Subframe must be used exclusively for one of these two purposes. Whereas in 5G NR, each symbol within the SLOT can be configured in a variety of ways. Theoretically, there would be almost an infinite number of possible combinations, but currently, only the combinations shown in Table 5.13 are the valid ones and are available for real operations.

Finally, it is important to evaluate the correct use of numerology, or carrier spacing, to identify when or under what conditions a configuration should or can be used. Figure 5.5 illustrates the various conditions of use.

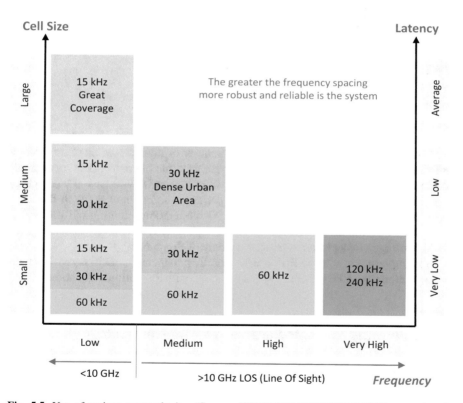

Fig. 5.5 Use of various numerologies. (Source: KEYSIGHT TECHNOLOGIES), reproduced with permission

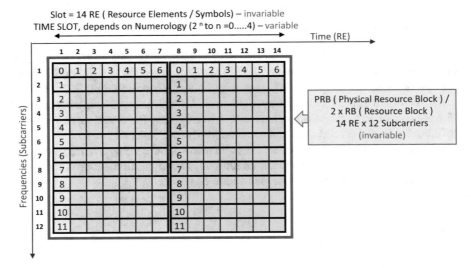

Fig. 5.6 Setting up a PRB – Physical Resource Block

Summary of Modulations Used in LTE and 5G

We have just analyzed how and what the 5G modulation schemes are. The complexity has increased proportionally with the number of alternatives that numerology provides. The complexity also results from the fact that 5G must maintain compatibility with LTE, which is the equivalent within numerology to "$n = 0$". To clarify any doubts on the subject, we decided to prepare a short summary with the main concepts that characterize both LTE and 5G modulation, pointing out the differences between the two technologies.

To fix these concepts, we will need the help of Fig. 5.6.

It is important to start by listing which of the parameters remain invariable with the evolution from LTE to 5G and which ones vary with the institution of numerology.

Invariable and Variable Parameters with Numerology

The first item, that remains *invariant, is the Frame dimension* that remains 10 ms. The *Subframe*, by consequence, also remains *invariant*, and equal to 1 ms.

The 4G slot is composed of seven symbols and its duration (Time SLOT) is 0.5 ms. In 5G, the SLOT is composed of 14 symbols and its duration is variable, as shown in Fig. 5.1.

The 4G *Resource Block* has a similar format, but not equal, to the 5G *Physical Resource Block*. In fact, one Physical Resource Block is equivalent to two Resource Blocks. While a Resource Block is a matrix formed by 7 symbols and 12 adjacent subcarriers, the Physical Resource Block is formed by 14 symbols and the same 12 subcarriers. While a Resource Block is formed by 84 Resource Elements, the Physical Resource Block is formed by *168 Resource Elements.*

A relevant factor in the evolution of LTE to 5G was the fact that it is possible to allocate downlink and uplink content indistinctly to each Resource Element of each Slot, as shown in Table 5.13. This was not possible in LTE and means a great advance for the reduction of latency, which is no longer plastered in the LTE TIME SLOT, which is 0.5 ms, because in LTE all Resource Elements of a Slot should be allocated to the up or downlink exclusively.

The main factor that varied between 4G and 5G generation was that the *number of subcarriers* used could be either 2048 or 4096. Although this factor does not depend on numerology, it is a new alternative.

What really varied with numerology was the *carrier spacing*, which is 15 kHz for LTE, and for $n = 0$. In 5G it could go up to 30, 60, 120, or 240 kHz, or even, in the future, go up to 480 kHz.

What are the consequences of this evolution?

- The *TIME SLOT*, i.e., the time duration of a slot (*Slot Duration*) follows the spacing between subcarriers. If this time is 1 ms for LTE and for $n = 0$ in 5G, for $n = 1$ is 0.5 ms, for $n = 2$ is 0.25 ms, for $n = 3$ is 0.125 ms, while for $n = 4$ is 0.0625 ms. With this, *it is possible to accommodate more Slots in the same Frame.* This makes all the difference in terms of throughput, because the number of Subframes, goes from 1 Slot per Subframe, reaching 16 Slots per Subframe for $n = 4$, as shown in Fig. 5.1. That is, the capacity is multiplied by a factor equal to 16 times. In the case of $n = 5$, this factor would increase to 32.

- With the decrease of the TIME SLOT width, the sampling time, or *Sampling Rate*, which is exactly the time that the system must identify the phase and amplitude changes suffered by the modulated signal, is much more critical. It becomes even more critical with the adoption of 4096 subcarriers instead of the 2048 used in LTE. If in LTE, with 2048 carriers the Sampling Time (Ts) is 32,552 ns, in 5G the Sampling Time (Tc) can reach 1017 ns, as shown in Table 5.5. This represents a huge advance in electronic circuits that detect phase and amplitude changes.

All tables and figures show, exhaustively, the changes in the other parameters that occurred with the evolution of LTE to 5G. This summary has been organized to highlight those that we believe are the most significant changes and that conceptually are the most important to emphasize, and about which there should be no doubts.

Chapter 6
AAS – Advanced Antenna System: The MIMO, Massive MIMO, and Beamforming Antennas

In the previous chapter, it was shown that the maximum possible 5G transmission rate is no more than 2.8 Gbps, even using the maximum available radio spectrum band of 400 MHz and the maximum modulation level of 256-QAM. How would it be possible to reach 10 Gbps, 5G's goal? That's what we'll see next. It is the active, "intelligent" antennas that provide this miracle.

First, it is important to note that the transmit/receive antennas are an integral part of the radio link system, where the physical communication of this technology takes place. That is why it is characterized as the lower part of the first layer of the IP (Low Physical Layer) stack, as will be discussed in a later chapter. The radio plus antenna set is what characterizes a wireless system, since the rest of the system takes care of signal processing within the concept of high-speed IP network, allowing communication with the network management Core.

The antenna has gained additional importance since 4G, when it was realized that there was a vast field to be explored in order to make more efficient use of the electromagnetic spectrum frequencies, which is no longer a passive element of minor importance, but rather a fundamental element of the technology, especially when it starts to explore the still "virgin" part of the electromagnetic spectrum, i.e., the millimeter wave portion, which brings very specific challenges. To properly understand how an antenna works, it is worth telling a bit of its history.

In 1862, James Clerk Maxwell developed his famous theory, demonstrating, theoretically, that electric and magnetic fields propagate with the speed of light. He did more, proposed the idea that light is nothing more than a wave composed of electric and magnetic fields that propagate simultaneously and orthogonally; that is, one field propagates in a plane and the other, in a plane that forms an angle of 90°. This had already been proposed by Michael Faraday. In 1887, Heinrich Hertz, by pure chance, verified that electric energy could originate electromagnetic energy thanks to a device which "irradiated" or transformed an electric current into electromagnetic propagation. This device, so-called a transducer for its ability to transform, was called an antenna. The movement of electrons circulating in a piece of conductor wire can transform the energy generated by its oscillation into electromagnetic energy, which propagates with the speed of light, also an electromagnetic wave. In honor of Hertz's discovery, the magnitude that expresses any type of wave, cycles per second, was named Hertz (Hz).

In a simple way, two conducting wires, one charged with positive charges and the other with negative charges, form a field around them, as shown in Fig. 6.1. If these charges are static, nothing happens. But if they are dynamic, thanks to the circulation of an electric current, they can radiate an electromagnetic field around them.

For this irradiation to be better used, it is necessary that a resonance occurs, like what happens with the strings of a piano or guitar, with the air when blown by a flute, with a bell, or even with a crystal glass when touched, transforming mechanical energy into sound energy.

Thus, for an antenna to radiate, a "stationary wave" must occur, which is nothing more than a wave that has a particular resonant pattern of vibration. The concept is simple. There is always an opposition to the phenomenon (reactance). Only when this opposition is minimized, the phenomenon occurs effectively. It is necessary, therefore, that there is a tuning between the mechanical characteristics of the antenna and the frequency of the current flowing through it.

The interesting thing is that an antenna can both radiate and "tune" an irradiance and transform an electromagnetic field into an electric current.

The most common type of antenna is the dipole, as shown in Fig. 6.2. As the speed of propagation of a wave in metals – usually the material used to manufacture antennas – is slower than propagation in air/vacuum, the size of the elements constituting the half-wave dipole ($\lambda/2$), the most commonly used in practice, must be somewhat smaller than the corresponding half-wave propagating in free space.

It is worth to point out an important remark related to the use of antennas in cellular systems. We had said that the spectral band used by an operator should be compatible with the transmission frequency of the radio frequency system. Let's remember that in 4G the maximum band considered is 20 MHz for a transmission

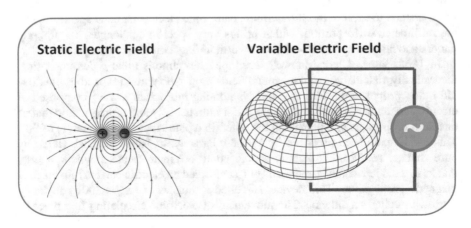

Fig. 6.1 Static electric field and dynamic electromagnetic field

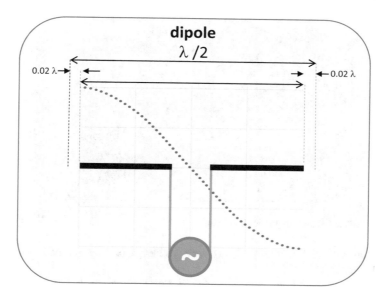

Fig. 6.2 Half-wave dipole antenna

frequency, in the most used case, 2.5 GHz. That is, we need to know how much 20 MHz represents in terms of amplitude variation around the main frequency, which in this case is 2.5 GHz and has a wavelength of $\lambda = 120$ mm. This variation represents only $+/- 0.004\lambda$. This is very important for antenna tuning.

In the case of 5G, let's analyze, for example, a relevant frequency of the FR1 band (<6 GHz), which is the 3.5 GHz ($\lambda = 85.7$ mm), and remember that the maximum band to be used is 100 MHz, variation of the transmission frequency equivalent to $+/- 0.01\lambda$. For higher frequencies, FR2 (>6 GHz), like 28 GHz ($\lambda = 10.7$ mm), the range is 400 MHz, which represents a "deviation" from the fundamental frequency; in this case, 28 GHz is only $\pm 0.007\lambda$. This ratio is essential for antenna sizing and should be negligible in order not to influence its tuning. As a rule, the bandwidth of an antenna, that is, the maximum amplitude of a band, should not exceed 10–15% of the value of the fundamental frequency, that is, its resonant frequency. We can observe that in all the examples shown above, the values for the various bands are much smaller than the recommended percentage.

Another relevant observation is that the two generated fields, the electric (E) and the magnetic (B), propagate orthogonally to each other, as proposed by Maxwell, and shown in Fig. 6.3.

The plane in which the electric field propagates defines its polarization, i.e., if it propagates in a horizontal plane, the polarization of the antenna is horizontal. If it propagates in a vertical plane, its polarization is vertical. It is worth remembering that it is possible to have simultaneous propagation of the two fields, maintaining only orthogonality between the two electric and magnetic fields without interference between them, as if two antennas were mounted at a 90° angle to each other.

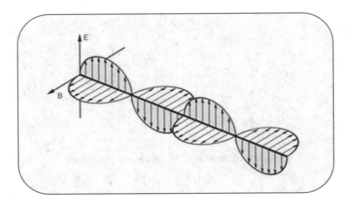

Fig. 6.3 Electric (E) and Magnetic (B) Fields

This concept is fundamental and was very well exploited by 5G, as it will be presented later.

An antenna is said to be isotropic when it radiates in all directions with the same amount of energy, but this is usually not desired. What matters is to focus the maximum possible energy in a particular direction, as is the case with the dipole presented above. The dipole has a radiation that resembles the shape of a donut and, therefore, can concentrate the energy in the region bounded by this shape. This represents a gain of energy, since it will be concentrated in a smaller region of space, while the isotropic antenna spreads energy evenly around it. This gain is usually defined always with respect to the isotropic antenna, and in the case of the dipole, this gain is 2.14 dBi. Regarding the unit dBi, we can say that dB is measured in logarithmic scale and compares very different magnitudes, being equivalent to 10 log (Power Radiated by Dipole/Power Radiated by Omnidirectional Antenna). The "i" indicates that the gain has as reference the isotropic or omnidirectional antenna.

Even the dipole is not ideal for most applications, where it is expected a higher concentration of energy in smaller regions. For this purpose, reflectors are used, which allow concentrating the energy in much smaller spaces, such as those shown in Fig. 6.4. The best known is the parabolic reflector, widely used in the reception of satellite signals. The larger the reflector, the greater the concentration of energy. However, this concentration is not always perfect, and some lateral "leakage" may occur. The main beam is called the main lobe, and the others are called secondary lobes. A dipole antenna with a simple reflector can achieve a gain of 3 dB only by concentrating the beam in one direction. A parabolic antenna can have gains in the order of tens of dB, reaching up to 50 dB, with a beamwidth of about 1°.

An antenna is characterized by the following factors:

- Its *gain*, that is, how well it can concentrate its beam in each direction with respect to the Primary Lobe, since secondary ones are undesirable.

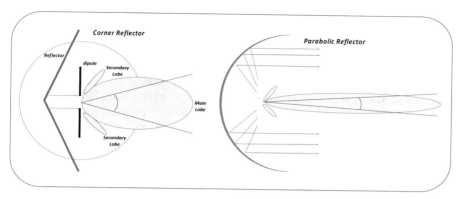

Fig. 6.4 Use of reflectors

- Its *directivity*, that is, the opening angle of its beam. This angle is calculated with respect to the half-power values (3 dB) presented in its *irradiation diagram,* which is exactly Fig. 6.4. The irradiation diagram is a three-dimensional figure, shaped like an ellipsoid of revolution, resembling a cigar.
- Its *polarity*, horizontal or vertical, allows double use of the spectrum within the same physical space, using two antennas orthogonal to each other.

The directivity of the antenna is very important, but it does not guarantee that the signal reaches its goal, just because along the way many things can happen. The main factors to consider are:

- The electromagnetic signal is always attenuated, and the attenuation of radio energy between two antennas is known as Free-Space Path Loss (FSPL), calculated by the formula FSPL = Dt × Dr × $(4\pi d/\lambda)^2$, derived from Friis' formula, in which (Dt) is the directivity of the transmitting antenna, (Dr) is the directivity of the receiving antenna, (d) is the distance and (λ) is the wavelength. (Note: In the case of two isotropic antennas, we have Dt = Dr = 1, so FSPL= $(4\pi d/\lambda)^2$). Attenuation can occur due to atmospheric variations, which can be random with minimal duration, or last for longer periods. These are called Large-Scale Fading or Small-Scale Fading (long or fast duration fading), which cause a higher signal attenuation. They can be selective (selective fading), affecting a small portion of the spectrum, or not. They are almost always unpredictable and may be caused by solar influence.
- Obstruction by walls, mirror glass, etc.
- Reflection, or multipath, is caused by buildings, constructions, or even vehicles passing in its path. Multipath can be constructive or destructive, as we will see later.
- Diffraction occurs when the signal encounters some obstacle in its path that causes the scattering of its beam.

Multi-path can be used in "intelligent" systems, because in-phase signals add up. But, if they are out of phase, they can cancel each other. This effect can be beneficial and needs to be addressed, because it allows introducing the concept of *multi-propagation*, which is a synonym of multipath, but it is used to make a given signal travel through different paths in a constructive way, as shown in Fig. 6.5.

In this case, the direct view of access to the RX receiver is obstructed, but some signals can be reflected, as in a mirror, so that it is possible to make the signal overcome the obstacle. It would be enough for a signal to be reflected. But let's take the opportunity to imagine that two signals reach the desired destination. The important thing is that the two paths taken, as shown in the figure, are equivalent to the same distance. In this case, the two signals would arrive "in phase" and would add up. If they are out of phase, the sum of the two could be destructive, but there would still be a solution if the signals leave the transmitter with a phase offset to compensate for the path difference. This would require an "intelligent" system, a quality that 5G has plenty of.

The concept of multipath is quite old in telecommunications, even in the days of analog systems. The multipath was part of systems that operated with Space and Frequency Diversity. The idea was to send the same information (signal) through different paths. Thus, more than one transmitting and receiving antenna was used, in the expectation that if one path was obstructed, the other would allow the transmission without major problems. This technique is called Spatial Diversity. The farther apart the antennas were from each other, the less likely the two paths would be to suffer interference. The same concept was valid for frequency diversity in which information was carried by two different carriers. Also, in this case, the more distant the two frequencies were, the better, because the smaller the chance that the two carriers would suffer the same fading. Such systems are not very efficient, because the received signals traveled through different paths, and in the reception only the signal that had the best conditions would be used and the others would be neglected, resulting in waste of spectrum. The technological evolution allowed much improvement in the use of resources and signal quality.

In previous cellular technologies, especially GSM, the use of the same frequency in nearby cells invariably resulted in the occurrence of interference and led to a huge loss of spectral efficiency. Spectrum is a scarce resource and must be used very well. The various cells used different portions of the allowed band and the reuse of frequencies was always limited. This restricted network expansion. In the case of WCDMA, this limitation does not exist, since all cells use the entire available spectrum, and the system uses distinct codes for each communication. But WCDMA presents serious difficulties at cell borders due to low carrier-to-Interference-and-Noise Ratio (*CINR*), forcing the use of low modulation levels, which is reflected again in low spectrum efficiency.

In the case of LTE, which uses many densely grouped subcarriers, when there is a simultaneous request from several users, the total capacity must be shared, and this maximum capacity also depends on the propagation conditions. To improve communication, LTE started using MIMO (Multiple Input Multiple Output) antennas, which, together with OFDM technology, allowed to remarkably improve communication performance and minimize the probability of transmission errors by using space diversity and *STC* (Space Time Coding). In 4G, all signals are treated

Fig. 6.5 Overcoming obstacles

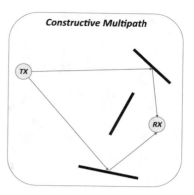

by means of a complex analysis process, so that a more robust communication can be extracted at the end.

MIMO technology uses a set of antennas constructed in such a way as to increase both the directivity and the total gain of the array. Note that now it is no longer a single isolated antenna, but a set of antennas, organized in order to constitute the so-called Gate Array. In this array, we can have, for example, four dipoles spaced generally ¼ of the wavelength ($\lambda/4$). The adjustment of this distance is important to maximize the concentration of energy and minimize unwanted secondary lobes.

For each dipole added to the array, there is a power gain as shown in Fig. 6.6. As we are always working on a logarithmic scale, the gain is 3 dB (10 log2). Therefore, in arrays of 8 antennas, the gain is 9 dB (10 log8). As the gain of a dipole is around 7 dBi, the total gain of the array can reach up to 16 dBi. In general, the gain of this array goes from 12 to 15 dBi.

Gate Array solves the problem of gain and directivity, which in turn minimizes interference from other cells. But it has the drawback of being static. To make it more effective, we would need to have several units covering the space bounded by the cell. The 5G brought something much more elaborate, the Phased Array, in which the beam is no longer static and becomes dynamic, as shown in Fig. 6.7. With this, we get to *Beamforming* technology, which allows directing the beam in the space covered by the cell.

The innovation consists of the possibility of compensating the differences in phase and amplitude so that the beam can be directed (Steerability) to the position in which it allows the best signal reception. And more: if the user is moving, the beam can follow the displacement, allowing maintaining the maximum gain of the antenna dynamically, in the desired direction. By adjusting the phase and amplitude, it is possible to ensure that the various signals generated by the antenna array add up so that constructive interference occurs in the direction of interest.

We also observe that, with the beam moving, it is possible to avoid undesirable interferences, a fundamental point for the use of high-level modulation.

Fig. 6.6 Gain of a gate
array

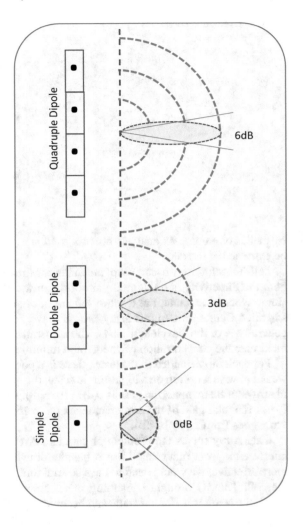

In 5G, the antenna array can hold several subarrays or subsets of antennas (with 2, 4, or 8 RF units), as shown in Fig. 6.8, in which the elements are used together to generate beamforming. In 5G subarrays, the dipoles are used in both polarities, thus doubling the transmission capacity.

As already presented, the larger the number of dipoles, as shown in Fig. 6.9, the higher the gain and the higher the directivity, which in this new case counts on two sets of antennas mounted orthogonally, taking advantage of the two polarization planes.

The antenna array can operate in several ways, as shown in Fig. 6.10.

In these three cases, a total of 8 antennas are used in each polarization, resulting in a total gain equal for all cases. In the first case, we see that each group of 2

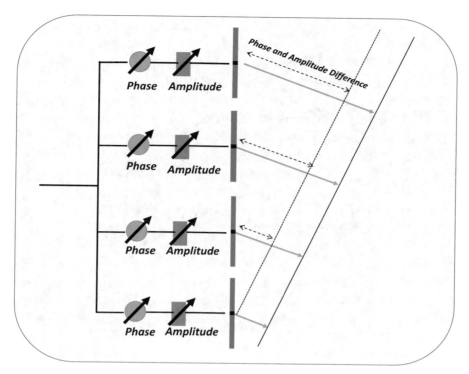

Fig. 6.7 Phased array

antennas (2×1) is dedicated to a single beam, allowing to serve more users simultaneously, but the gain of each beam is lower, covering a smaller radius, typical of densely populated (urban) regions. In the second case, 4 antennas are dedicated to each beam, making it possible to reach greater distances, but serving fewer users, typical situation of the suburban region. In the third case, all antennas are dedicated to a single beam, which has a larger coverage radius, being indicated for a less populated area, typically rural.

These panels can be arranged to work with groups of up to eight antennas, as shown in Fig. 6.11, mounted next to each other, resulting in an array of $8 \times 8 = 64$ antennas. This configuration is known as *massive MIMO*, and these panels can work in the following modes:

It is also important to evaluate how the RF transmitters/receivers of these antennas are connected. This is shown in Fig. 6.12.

It is important to note that we are actually working with several radio systems, each feeding an antenna positioned in a certain polarization. In Fig. 6.12, we have two sets of radios each feeding orthogonal (cross-polarization) antennas.

Therefore, a new alternative arises, sending more than one stream to the same user, allowing the user to receive multiple streams simultaneously. This greatly

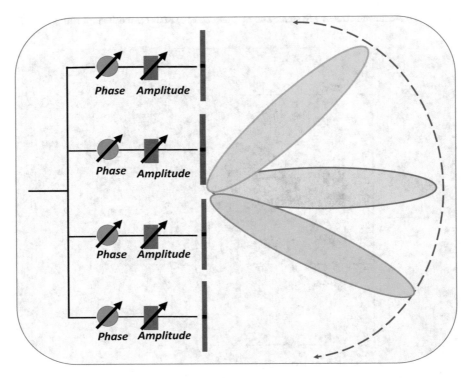

Fig. 6.8 Beamforming antenna

Fig. 6.9 Gain variation as a function of the number of dipoles

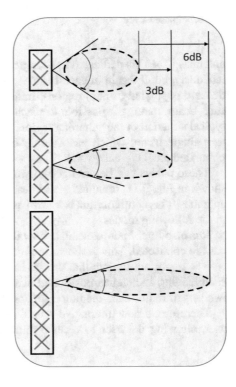

increases the possibility of providing a much higher throughput to a single user. Each stream is called a *layer*. Multiple layers can therefore be allocated to a single user. The maximum number of layers allocated to a single user does not depend on the transmitter, but on the number of antennas at the receiver side, because each antenna must receive a stream, and the big limitation, in this case, is the user equipment (UE). The most usual, presently, is to work with up to four layers. The number of layers available in a device is called *rank* and depends on the number of radio streams that the equipment can process simultaneously. So, theoretically, a single user could receive up to 4×2.8 Gbps $= 11.2$ Gbps, *reaching the goal to be achieved by 5G (>10 Gbps)*. The number of layers can be expected to reach 8, increasing the Transmission Rate to more than 20 Gbps.

It is important to remember that the propagation of the layers can occur in different polarizations and even different paths, taking advantage of some point where one of the beams can be reflected.

In the image in Fig. 6.13, we can see the case of SU-MIMO (Single User – MIMO), in which a single user is being benefited by more than one layer. But this same facility can also be applied to multiple users. In this case, the denomination becomes MU-MIMO (Multi-User – MIMO). The figure also shows how beam selectivity minimizes possible interference.

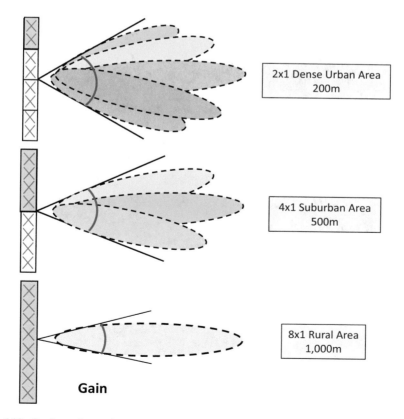

Fig. 6.10 Configuration options of an antenna array

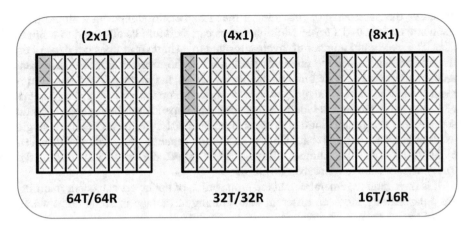

Fig. 6.11 Massive MIMO settings

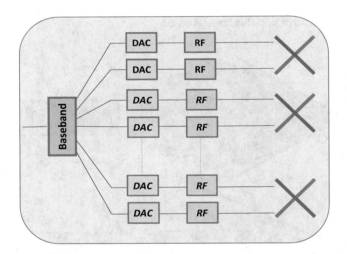

Fig. 6.12 RF connection × antennas

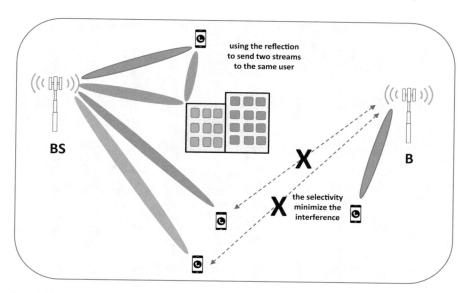

Fig. 6.13 Use of multiple layers

Chapter 7
The Different Architectures Used in 1G, 2G, 3G, 4G, and 5G Networks

Architecture is always related to the processing flow, determining the speed, form, and capacity that the system will have to treat the transmission of a signal. The structure defines what is expected from what is built. By comparing the various architectures, we can observe the evolution of the different generations of mobile phones over the more than 40 years of existence of mobile networks. This is the subject of this chapter.

The general concept of a network consists of three parts: The Air Interface, Access Network, and Core Network, as shown in Fig. 7.1.

We also note that, from the *CORE*, it is possible to interconnect the system with other mobile phone networks, landline telephony, and even with packet networks and the internet, integrating the cellular network with the global telecommunications system.

The air interface has two basic components. At one end is the user equipment, which in GSM is called *MS* (Mobile Station) and in later generations receives the name of *UE* (User Equipment), usually represented by a mobile phone, but it can also be a laptop, a tablet, or any other device that has an appropriate interface to access the cellular network. At the other end, we have what can be generically called a *Radio Base Station (RBS)* or Base Station (*BS*), a name used in the first generation, but which over the years has been changed in accordance with technological changes. In 2G, it was called *BTS* (Base Transceiver Station); in 3G, *Node B*; and in 4G, *eNodeB* (evolved Node B). In 5G, it is called *gNode B* (next generation Node B).

The *Access Network* gathers and consolidates all connections (accesses) of a network composed by several BSs and serves as a connection interface between the user's equipment and the *Network Core (CORE)*, where the information, signaling, and calls, are effectively processed. The Access Network receives the name *GERAN* (GSM/EDGE Radio Access Network) in 2G, *UTRAN* (UMTS Terrestrial Radio Access Network) in 3G, *RAN* (Radio Access Network) in 4G, and *NG-RAN* (Next Generation Radio Access Network) in 5G.

© The Author(s), under exclusive license to Springer Nature Switzerland AG 2023
J. L. Frauendorf, É. Almeida de Souza, *The Architectural and Technological Revolution of 5G*, https://doi.org/10.1007/978-3-031-10650-7_7

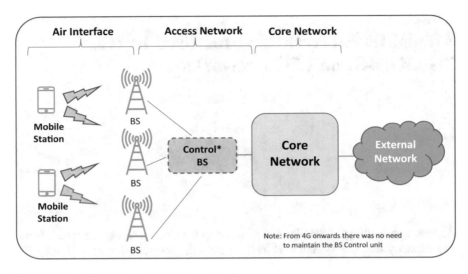

Fig. 7.1 Simplified scheme of a cellular network

The dialogue between the user terminal and the Core Network obeys a determined protocol, which is the language used by the network in processing the information and which will be detailed in the following chapter. But we can already anticipate that over the years, much of the radio interface functions have been automated and transformed into software, while the hardware has benefited from virtualization and new technologies, which have allowed increased efficiency and reduced the size of components so that today a set of highly sophisticated chips is responsible for transmitting and receiving signals.

The architecture of the first three generations shows well its origin, focusing on the telephone service. Only as of 4G did the focus shift to access to the IP network. Figures 7.2, 7.3, and 7.4 show the evolution of the first three generations.

It can be noted that the architecture of the first three generations is quite similar as far as the telephone service is concerned, being composed of a unit that does the signal transmission and reception processing involved in the conventional telephone communications (BS) and a unit that aggregates and controls the BSs traffic. According to the cellular generation, we have:

1G *MSC* (Mobile Switching Center), which aggregates BS (Base Station) traffic and interfaces directly with the Public Switched Telephone Network (PSTN). Note that first-generation mobile telephony only allowed voice communication and used the same Core Network as the fixed-line network, represented by the MSC.

2G and 2.5G *BSC* (Base Station Controller) aggregates and controls the traffic of the BTSs and passes the telephone traffic to the *MSC*. The MSC is part of the Circuit Switching (CS), which is the GSM Core Network. Note that the MSC, a legacy technology of the public switched telephone network, is still the main element of

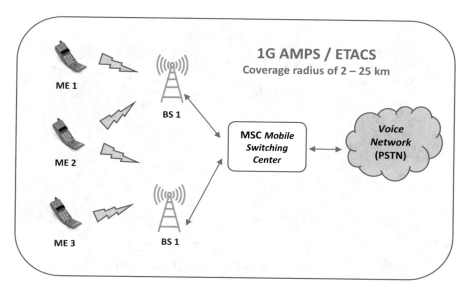

Fig. 7.2 First generation cellular system: AMPS/ETACS

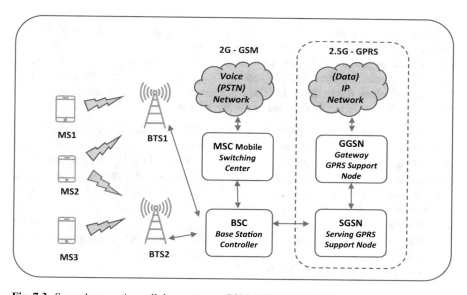

Fig. 7.3 Second generation cellular systems – GSM (2G)/GPRS (2.5G)

the GSM Core. Circuit Switching (CS) allows 2G to be connected to the PSTN. GSM is still a system focused on voice transmission. In 2.5G generation, the adoption of GPRS architecture allows interconnection with packet network (X.25 and Frame Relay), low-speed data transmission (9.6 to 171 Kbps), and *SMS* (Short Message Service) text messages.

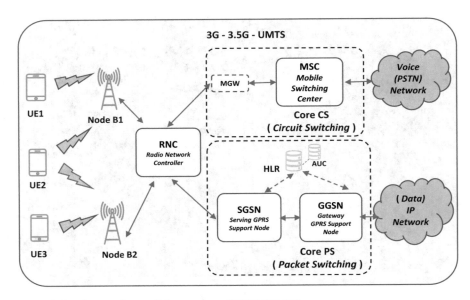

Fig. 7.4 Third generations cellular systems – UMTS (3/3.5G)

3G and 3.5G In this evolution phase, the BSC is replaced by the RNC (Radio Network Controller), which promotes traffic aggregation and controls the NodeBs. Data traffic is processed by SGSN and GGSN and then directed to PS (Packet Switching), and telephone traffic is sent to CS (Circuit Switching). There, the MSC is used to process the voice calls. The PS interconnects to the packet/internet network, and the CS interconnects 3G to the legacy networks.

The BSC interconnects the BTSs with the MSC in the Core CS, controls the radio functions in the GSM network, participates in resource management, and manages the handover with the help of the MSC. In turn, in 3G (UTRAN), the RNC is responsible for controlling the NodeBs, controlling the radio resources in its domain, and doing mobility management besides being the intermediary component between the NodeB and the Core (CS and PS). In short, the BSC and the RNC have similar functions, but are differentiated by technological advances that were introduced in each new version of the mobile phone system. Basically, the RNC is more intelligent, for example, the RNC can manage handovers autonomously, without involving the MSC or the SGSN.

But with the introduction of data traffic, it is necessary to introduce new elements and features. One of them, the *PCU* (Packet Control Unit), was added (overlayed) to the *BSC* (Base Station Controller) in 2G to turn it into 2.5G, allowing data transport. This device diverts data traffic, separating it from phone traffic. Thus, our well-known *GSM* (Global System for Mobile Communications) is also known as *GPRS* (General Packet Radio Service). The *RNC* (Radio Network Controller) of 3G already had, since its conception, this fundamental module, which allowed supporting the manipulation of data packets. Moreover, the Core of the UMTS system (3G) received the addition of *PS* (Packet Switching), allowed the connection with the Packet Network. Thus, the following units emerged:

SGSN (Serving GPRS Support Node), which is responsible, among other things, for identifying the user's location, authentication and generate a special encryption algorithm called GEA (GPRS Encryption Algorithm) and do the routing of traffic packets.

GGSN – Gateway GPRS (General Packet Radio Service) Support Node provides the interconnection of the cellular network with the internet or even with private networks through gateways or ports that interconnect the various networks. It can also be seen as a combination of gateway, router, and firewall because it serves as an interface between two environments, the internal and external.

It can be seen, therefore, that the architecture, in essence, has not changed over the first three generations regarding telephone traffic (voice), which, by the way, continued to be treated in the same standard until the emergence of 4G, which in its complete form, i.e., LTE plus IMS (IP Multimedia Subsystem), known as VoLTE (Voice over LTE) system, allows voice and data processing in a fully IP environment. To make 4G commercially viable, the operator was given the opportunity to acquire only the LTE system similarly to 3G's Packet Switching, which allowed the user to access the packet/internet network but had no ability to process phone calls. In this option, when it received a voice call request, the 4G simply transferred it to the 3G, in which the call was processed. After the call was disconnected, the system took care of returning the link to the 4G LTE and reconnecting the user to the packet network. We observed that, despite the investment required to install the IMS, the full VoLTE system offers great advantages, including the end-to-end treatment of voice and data calls within the 4G network, traffic relief over the 3G network, and access to more elaborate services offered by 4G.

Radio Access Network (RAN) Architecture

Before we go into detail about the 4G LTE architecture, we must open a parenthesis to make some important considerations regarding how the concept of cell dimensions evolved and how this resulted in the need for changes to the architecture in the RAN due to the growing demand for traffic and the constant search for improving quality of service, which includes extending coverage into shadow areas, where the radio signal previously did not reach.

The change in the concept of cell size was considered when launching 4G cells, which already use frequencies close to the 3 GHz band. The need to provide better coverage for the provision of services was evident. With this, 5G brought with its new concepts of area sizing. We will advance in this topic to show the differences between previous generations and 5G.

The transmission range of a cell depends on several factors, such as the nominal power of the transmitter, frequency, technology used, local geographic factors – such as the existence or absence of obstacles – antenna height and even weather conditions. Traditionally, 2G and 3G technologies, which operate at frequencies below 3 GHz, reaching a maximum of 2 GHz, work with macrocells that, in line-of-sight, can reach a few kilometers.

4G occupies a higher band, 2.6 GHz, and works with macrocells, but with reduced coverage. On the other hand, 5G was developed for a high density of connections (one million connections per km^2) and can work with several frequencies, including millimeter waves, which have a much smaller coverage area, but with higher data throughput. For this reason, 5G cells are dimensioned for small coverage areas and called Small Cells. Operating at higher frequencies, the coverage radius is around 250–300 m, or even smaller, when using millimeter waves.

5G operates with a radius of the order of 1–2 km, using higher powers, which can reach 50 W, and which will be primarily intended to serve suburban areas. To serve more dense areas, *Microcells* were also created, with a coverage radius that can vary from 250 m to 1 km, intended to serve a large number of users, operating with powers of the order of 5–10 W. The *Picocells*, on the other hand, aim to serve shadow areas with coverage radius varying from 100 m to 300 m, working with powers of the order of 2 W. Finally, to attend the areas with intense and localized traffic demand, we have the *Femtocells*, that cover an area with radius from 10 m to 50 m and operate with powers of about 200 mW. Femtocells are actually of two types: those for commercial use, which can serve more users, and residential, which serve a smaller number of people. It should also be noted that 5G is designed to allow the integration of different types of access. For this reason, it is expected that, in addition to 5G Small Cells, solutions such as Wi-Fi will also continue to be widely used in the future, especially indoors, and that even 4G macrocells will remain active for several years.

This set of alternatives is available to compose what is called a *RAN* (Radio Access Network) or, by extension in 5G, NG-RAN (New Generation RAN). A RAN is characterized by a large number of cells, each employing the appropriate equipment to meet its specific characteristic, aiming to provide the greatest possible coverage.

Initially, see Fig. 7.5, the conventional base station can be divided into three parts: the *Antenna*, which is the radiating element, the *RRH/RRU* (Remote Radio Heads/Remote Radio Unit), which constitutes the radio frequency modulation (carrier) part, and which connects to a digital processing unit (OFDM modulation), completing the characteristic functions of a base station. This second unit is called a *BBU* (Base Band Unit). The BBU, in turn, connects to the Core Network. This configuration is called *D-RAN* (Distributed Radio Access Network) or, in rare cases, Decentralized Radio Access Network as opposed to Centralized RAN (*C-RAN*), whose important function will be discussed below. The connection between the antenna and the RRU/RRH is made by coaxial cable whose length should be short to avoid large losses by attenuation.

It is important to note that this architecture is typical of technology in which the antenna is considered a passive device. Everything changes with the introduction of active MIMO/Beamforming antennas. The connection between RRU and BBU is usually made by optical fiber, because it is a long distance, and follows the *CPRI*

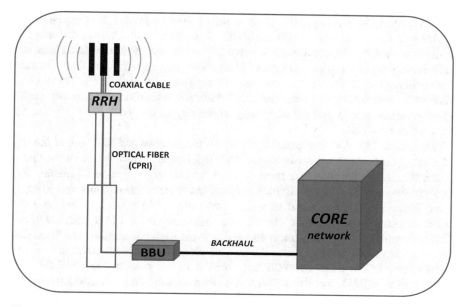

Fig. 7.5 Traditional D-RAN architecture

(Common Public Radio Interface) protocol, or its latest version, *eCPRI* (enhanced Common Public Radio Interface). The connection between the BBU and the CORE is made through the Backhaul usually using a fiber-optic network, but optionally also by a point-to-point microwave link or by another communication solution convenient to the system, using appropriate protocols.

In the case of 5G, the use of fiber is almost mandatory due to the volume of traffic. Just to give an idea, when using three massive MIMO antennas with 16 elements per sector, with each antenna serving each of the three sectors of a cell, the traffic volume of the backhaul could be around 100 Gbps. Besides the volume of traffic, latency is an important parameter in 5G, demanding a Backhaul that meets the timing criteria of this technology.

This configuration, shown in Fig. 7.5, initially used in 4G, implies that the BBU, installed near the antenna, requires the construction of a shelter equipped with air conditioning and power supply, within a standardized model. All this makes the operation of the cells more expensive and complicated.

As traffic demand has increased, the amount of radio equipment required to serve a coverage area has increased and evolved into a model where the BBUs are all concentrated to serve multiple antennas in a certain region, as shown in Fig. 7.6. This format is also known as *C-RAN* (Centralized Radio Access Network). The main facility (BBU Hotel) acts as a "hotel" where several BBUs are housed.

This architecture highlighted an interesting aspect. The BBUs, although close to each other, worked as independent units, and this generated almost a dispute between them to serve users at the edges, regions where there is an overlapping performance of the various cells. Then came another evolution: to concentrate all BBUs in a single processor with the ability to serve several cells simultaneously, forming a pool. As the processing is unique, it eliminates the competition between cells in serving users, who can be served by more than one of them, thus allowing the optimization of resources, especially in regions with high traffic fluctuation, such as those exclusively commercial, and stadiums, whose service capacity varies greatly during certain periods of the day or the day of the week itself. Figure 7.7 shows this evolution.

The initial, simpler configuration, in which the antenna and RRU are at the top of a tower or mast and are connected to the BBU, housed at the bottom of the tower preferably by optical fiber at a short distance, of the order of 20 to 50 meters, has evolved into a system in which the RRU remains close to the antenna, but is separated from the BBU by long distances, of the order of 15 to 20 km, and both are interconnected by optical fiber. The use of coaxial cable at the CPRI interface of the C-RAN architecture is not recommended due to the high attenuation. The RRU can be integrated to the antenna, as is the case of Picocells and Femtocells, due to the reduced size of the set and because they work with low powers. However, this can also occur in MIMO, massive MIMO, and Beamforming configurations. The most appropriate way to interconnect high volumes of traffic is using optical fiber, which can even be shared since each connection is made by a beam of light working at a specific wavelength. In this way, different wavelengths can be multiplexed sharing the same fiber. The ideal is to work with an optical ring, where if there is an interruption at any stretch, the transmission still remains "on air" without interruption. This is possible simply by reversing the direction of transmission/reception of the light beams.

The C-RAN architecture aims to reduce installation and maintenance costs while operating more efficiently, enabling the processing of a higher volume of traffic, as

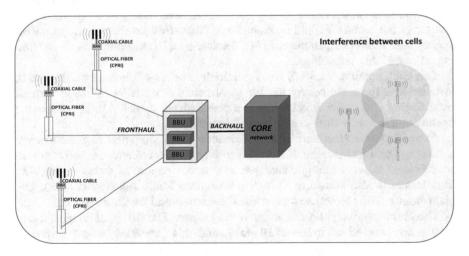

Fig. 7.6 C-RAN BBU hotel architecture

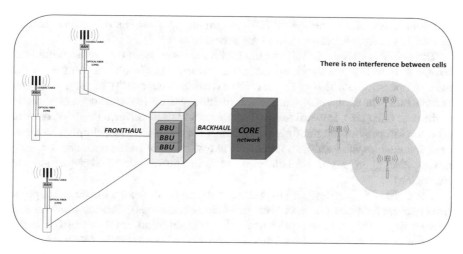

Fig. 7.7 C-RAN pool architecture

well as increasing network security and facilitating system expansion whenever demand requires it. The term "Cloud RAN" (also expressed as C-RAN), sometimes so named, indicates that the RAN makes extensive use of cloud computing resources, such as the virtualization of network functions and the dynamic adjustment of capacity and resources according to traffic.

We have shown how the architecture was conceived without entering the dynamic operation of the system, and just understanding who does what. Next, we will verify which functions are performed by each block. In the next chapter, we will evaluate how communication is done, and the dialogue between the units. We will start by analyzing what happens in 4G to then understand how the evolution to 5G occurred. How does this happen?

In 4G LTE, when the user equipment (UE) is turned on, a communication channel is established with eNodeB over the air, and eNodeB, in turn, establishes another communication channel for sending control signals to the *EPC Core* (Evolved Packet Core). Through this channel, it is possible to authenticate users on the network, record and monitor their movement within the coverage area, and control the handover and the state of the user's terminal. The portion of the Core responsible for these functions is called *Control Plane*, as explained in detail below. The portion of the Core responsible for interconnecting the user and the packet/internet network is called the *User Plane*.

After user registration, eNodeB creates a "tunnel" through which the user terminal can exchange data directly with the packet network using 3GPP terminology, a *Bearer Service*. With each new application accessed by the user, a new data bearer is created between the user terminal and the packet network. When "logging in" to the packet network, the UE receives an IP address, which is assigned to it provisionally until the device disconnects from the network. As such, the eNodeB allows interconnection of the user device (UE) to both the Control Plane to send signaling and the User Plane to gain access to the packet network.

The eNodeB stations coordinate the service to users when they move within the coverage area of the network operator. The role of the eNodeB is to observe the rules of operation of the network in order to maintain the quality of services (QoS) and can make the decision to send the information packets (data and control) directly to the user unit (UE), forward them to another eNodeB to perform this task, or even send the information by multiple paths. Thus, the eNodeB is not simply a component that enables the handover. It can also play the role of traffic optimization, promoting the aggregation or balancing of this traffic (Link Aggregation/Link Balancing).

The Core EPC was designed to operate in a pure IP environment and can support real-time applications (such as VoIP or video applications), besides conventional data traffic, strictly following the IP Protocol. In addition, the Core EPC has the function of connecting the local operator network with external networks, such as legacy networks of other generations of mobile telephony and networks from other operators, or even making the connection with the packet/internet network, ensuring that the interconnection occurs within specific security and quality criteria.

Formally, the *EPS* (Evolve Packet System) is composed of the 4G access network, which is called *E-UTRAN* (Evolved Universal Terrestrial Radio Access Network) or *LTE* (Long Term Evolution), *and* the *EPC* (Evolved Packet Core). This network is composed of the user terminal (*UE*), which communicates with the base station *eNodeB* (evolved NodeB) via the air interface. The eNodeB, in turn, communicates with the network Core, EPC. This architecture was named *SAE* (System Architecture Evolution), as shown in Figs. 7.8 and 7.9.

In the representation of Fig. 7.9, we see that eNodeBs located in the same area communicate to each other through the interface called *X2*. This is an important point in the implementation of some of the network functions. The eNodeBs communicate with the Core EPC through the *S1* interface, while the Core EPCs communicate through the *S10* interface. S1 interfaces can also be established between an eNodeB and other Core EPCs to speed up communication between all network components.

This new architecture was designed with the objective of distributing the intelligence among the various network elements, as well as increasing operational efficiency, aiming to accelerate the connection of users and reduce the transfer time (handover) from one eNodeB to another, when the user moves within the performance area of the various cells that make up the network, a fundamental factor for the quality of mobile communication.

Another key factor in this distributed architecture is that the scheduling of connections occurs through a dialog between the UE and the eNodeB, making communication much more effective. The Scheduler is responsible for ordering the communications. It is a key component to make efficient use of radio resources. The communication of the various users follows a logical sequence in order to optimize the transmission of each communication, allocating each content within the Frames and TTI (Transmission Time Interval), which in the case of LTE lasts 1 ms, and that

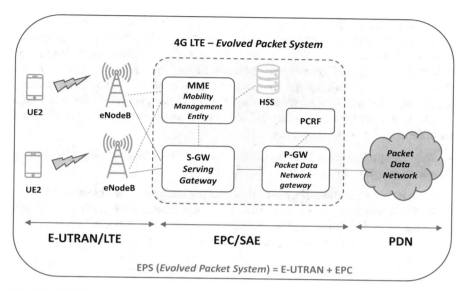

Fig. 7.8 4G LTE access network architecture – evolved packet system. E-UTRAN (evolved universal terrestrial radio access network) + EPC (evolved packet core)

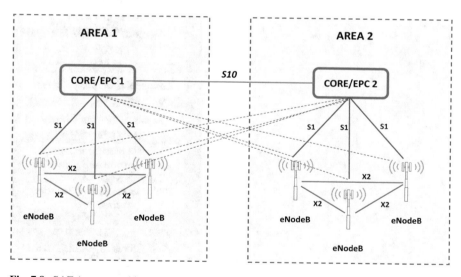

Fig. 7.9 SAE (system architecture evolution)

was discussed earlier. In 5G, to reduce latency, this period has been reduced and may vary from 1 ms to 125 µs, according to the adopted numerology.

Let's evaluate in more detail each of the blocks that make up E-UTRAN. What are the functions performed by *eNodeB*?

The eNodeB, as shown in Fig. 7.10, is the main and only interface to the user equipment (UE), it houses the main layers of the IP stack, the *PHY* (Physical Layer), the *MAC* (Medium Access Control Layer), the *RLC* (Radio Link Control Layer), the *PDCP* (Packet Data Convergence Protocol Layer), and *RRC* (Radio Resource Control). This topic will be the subject of our next chapter. For those not familiar with IP network architecture, we have organized a summary on this subject in Chap. 13. But before we get into the "layers" of the IP network, we should understand the main functions processed by eNodeB:

- *RRM* (Radio Resource Management) is the unit that manages the physical radio resources, performing the following functions:

  - *Dynamic Resource Allocation:* dynamic allocation of available resources.
  - *Admission Control:* access control (admission), authorizes, or not, to establish the tunneling (Bearer). The creation of *Bearers,* or "tunnels," allows direct access of the UEs to the packet network *PDN* (Packet Data Network), i.e.,

Fig. 7.10 Functions
performed by eNodeB

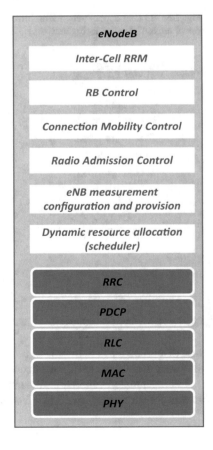

direct access to the internet. This connection is created, maintained, or broken under the control of Admission Control.

– *Mobility Management:* Manages the radio resources related to the connection to the UE in both Idle and Traffic mode.

- Selecting the EPC to which the UE will be connected via one of the eNodeB that make up the network, as shown in Fig. 7.9, allows it to connect to the chosen *MME* (Mobility Management Entity), as will be seen below.
- *Routing:* forwarding control signaling data to the MME and content data to the *SGW* (Serving Gateway). These elements will be better detailed below.
- *Scheduling Enforcement:* scheduling aims to ensure *QoS* (Quality of Service), i.e., transmission quality management, allocating resources according to the *SLA* (Service Level Agreement), that is, according to the service contracted by the user.
- *Encryption/Decryption:* Encryption has the function of guaranteeing the integrity and secrecy of the air interface contents.
- *Header* Compression/Decompression: header compression/decompression is done to minimize the volume of information traversing the network.

We will now analyze the functions performed at the control center, the Core EPC. Figure 7.11 shows the units that make up the EPC and how they communicate. The EPC is considered a "flat" architecture, which allows the management of a large volume of data in an efficient, safe, and economical way.

The Core, or core of the LTE network, is composed of a set of three basic units (MME, SGW, and PGW) and some complementary units (HSS, AAA, and PCRF). Further detailing:

MME (Mobility Management Entity) Main control node for the LTE access network, it processes all the signaling exchanges between the UE and the EPC, being responsible for authenticating the user with the help of the HSS (Home Subscriber Server), as well as applying roaming restrictions to the UE, when they exist. The MME is responsible for the UE mobility control, including tracking and paging, and for locating the user equipment (UE) when in idle mode. The MME is involved in carrier activation and deactivation signaling and is responsible for selecting the SGW in the UE's initial connection (attach) process. In addition, it handles signaling through the NAS (Non-Access Stratum) and is responsible for encryption and protection of signaling integrity. It is also responsible for the generation and allocation of temporary identities for the UE (GUTI – Globally Unique Temporary Identifier). The MME also controls the mobility between LTE and 2G, 3G, and 5G networks. The MME participates in the handover processes in case the eNodeBs are not connected to each other through the X2 interfaces, or when an inter handover occurs, which involves the relocation to another Core Network. If the handover occurred between eNodeBs connected to the same Core, it would be called intra handover (handover within the same Core Network). The MME is normally connected to the HSS/AAA, which contains the entire subscriber database and provides this information to the MME when requested.

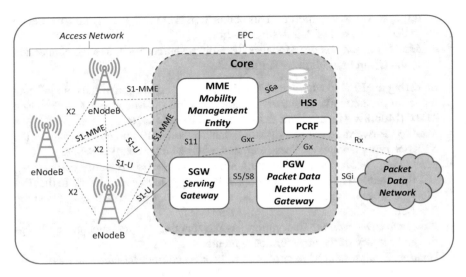

Fig. 7.11 LTE core architecture/EPC (evolved packet core)

AAA (Authentication, Authorization, and Accounting) Server that supports the MME in performing the authentication, authorization, and accounting functions. In the most modern models, the HSS has the AAA function incorporated.

HSS (Home Subscriber Server) Subscriber database, that concentrates all the users' information. It controls the parameters of the contracted services, validating the roaming conditions, and controls the "keys" related to encryption, assisting in the security of communications. It also monitors the user's geographical positioning within the network, continuously updating its position.

S-GW (Serving Gateway) All user IP packets are routed through the S-GW, which acts as the Gateway for the connection. It cooperates with the MME in the creation of bearers and assists in the P-GW selection process, which will be described next. Maintains the information about the bearers when the UE is in ECM-IDLE state (inoperative). Acts as a local mobility anchor in cases of handover between eNodeBs and collects administrative information related to charging for network usage in cases of roaming. It acts together with the P-GW in the implementation of the lawful interception function.

P-GW – Packet (Data Network) Gateway The P-GW is the mobile network's gateway to the internet, being the main router of IP Packets received or to be transmitted. A UE may be simultaneously connected to more than one P-GW, if the mobile network is connected to several external networks. It is responsible for assigning the IP address of each user on the network. It is interesting to note that the S-GW and the P-GW may be physically together, that is, within the same device, but each one exercising a distinct logical function. The P-GW can be used in the function of interception of communications authorized by law in which it acts with the S-GW. In networks where a Policy Server (PCRF) exists, the P-GW also acts as PCEF (Policy and Charging Enforcement Function), allowing the application of rules related to traffic control, filtering data packets, and supporting the system responsible for network policies and/or service billing. This process will be explained below.

PCRF (Policy and Charging Rules Function) The PCRF is a network policy server that, although considered a complementary unit in the Core EPC, with the constant increase in network traffic volume, has become increasingly important in establishing traffic control rules that can be applied online and allow managing and organizing traffic, helping to keep the network in good operating conditions and maintaining the quality of service (QoS). The PCRF can even make the charging of the services provided (CRF – Charging Rules Function), if the network does not have a specific charging system. In larger networks, traffic management can be complex. To avoid problems such as link congestion, service interruption due to traffic overload, and misuse of the contracted bandwidth (such as when a user purchases a voice service and uses that link to transmit video), among other problems that may affect service quality, operators may resort to a network policy server. In 4G, this function is implemented by PCRF.

PCEF (Policy and Charging Enforcement Function) Logical function, it is implemented through some network elements under the supervision of PCRF, as is the case of P-GW. It is responsible for ensuring that the rules established for data flow control are observed, ensuring quality of service (QoS).

It is important to remember that when 4G was conceived, the focus changed and data transport became the main objective of the cellular network, which ceased to be a simple "mobile telephony" network. Moreover, after a few years, it can be said that 4G was the embryo of 5G. For this reason, we will treat the two generations together, since at first 4G may share part of the existing structure with 5G in the NSA model (Non-Stand Alone), and probably in a near future the 4G will merge completely with 5G, as we shall see below. Note that Base Stations are called *eNodeB* (evolved NodeB) in LTE and *gNode B* (next-generation Node B) in 5G.

Below, we see Figs. 7.12, 7.13, and 7.14 which show the three stages of this evolution.

Although the Core EPC was designed for LTE, with the emergence of 5G a solution was devised to use the existing EPC infrastructure to support the introduction of this new technology. It was observed that certain devices or applications demanded a high traffic of messages related to signaling/control of the transmission, while other applications required a reduced volume of control messages. This is linked to the growing volume of access to video content, for example, while other applications have already indicated the need to create a more appropriate model, as in the case of communication between machines (M2M) or even the IoT (Internet of Things). The response did not last long and the concept of separating the exchange of control information, *Control Plane*, from the content information, *User Plane,* was conceived and named *CUPS* (Control and User Plane Separation), making it possible to scale networks according to the specific needs of the Control Plane, regardless of the User Plane, increasing network flexibility, optimizing investments, making traffic much more efficient and reducing traffic in the backhauls serving the network.

The separation of the two functions also occurs at the physical level, as we shall see below. The first separation was conceived as purely logical, i.e., the processing occurs in the same network device, which treats each function independently (Inline CUPS). Soon, it was realized that this was not satisfactory, and the functions should be separated, with a dedicated device for each function (Co-located CUPS) but located in the same data center. The evolution did not stop there. Since nothing prevents functions from being processed in geographically separated data centers, the Remote CUPS model was created. Figures 7.15, 7.16, and 7.17 show this evolution.

The gains are easy to enumerate: scalability of each function according to demand and, perhaps most importantly, the use of a Control Plane to manage multiple User Plans. In addition, the User Plane can be closer to the user, substantially decreasing the Latency Time, whose decrease is so desired in 5G and, concomitantly, decreasing traffic in backbones. This generated the concept of *Edge Computing*, in which processors are relocated to regions near the edges of the cells. This is especially useful for OTT (Over-The-Top) applications, such as Netflix and Prime Video, in which content servers can be closer to areas where the services are provided.

Following this concept, it is not difficult that a simple Control Plane, as shown above, can serve a metropolitan area in which there are several sets of SGW-U/

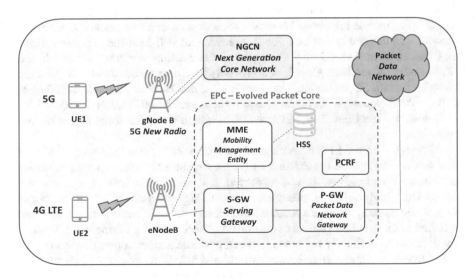

Fig. 7.12 4G and 5G independent architectures

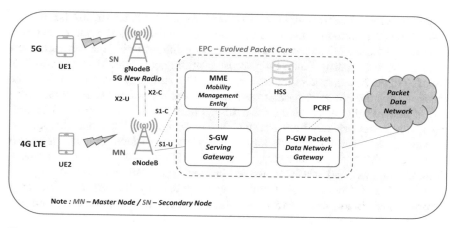

Fig. 7.13 5G NR – 3GPP (option 3) NSA (non-stand alone) architecture

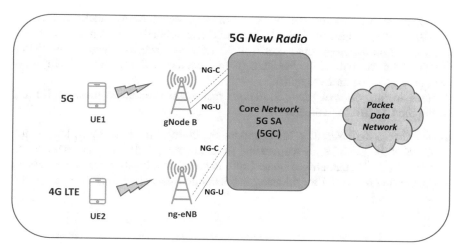

Fig. 7.14 5G-SA stand alone final architecture: 5G and 4G users

PGW-U distributed in about 10 areas, to which around 100 Base Radio Stations may be connected.

We will now understand how the LTE Core evolved to become *5GC* (5G Core). The first, called Reference Point Representation or Non-Roaming 5G System Architecture, follows the traditional architecture, inherited from the older technology, which is characterized by point-to-point connections between the various functions that make up the assembly. Figure 7.18 illustrates this alternative.

The second version, as shown in Fig. 7.19, is more current to meet the new demands, becomes, from this new generation, totally service-oriented and for this reason receives the name of *SBA* (Service Based Architecture).

We must remember that any access network must be able to perform basic functions, such as communicating with the user's equipment and providing access to other networks, storing data related user's identity and regarding the contracted services, providing security, controlling access admitting only authorized users and, in the case of mobile networks, controlling mobility. In the new 5G network, some functions that did not exist in other generations have been added to allow the adoption of new paradigms and to meet the differentiation of services provided according to the necessary requirements, which has resulted in "network slicing." As already discussed in the introductory chapters, the main requirements are *eMBB* (enhanced Mobile Broadband), *URLLC* (Ultra-Reliable Low Latency Communications), and *mMTC* (massive Machine Type Communications).

In the 5G architecture, it is extremely important that any function, provided it has the proper permission, can have access to the services of another function. Therefore, the 5G *NFs* (Network Functions) use exclusively service-oriented interfaces in their interactions. These interactions occur on a communication bus, similar to the buses used in computers. This is not a coincidence, since this new architecture is implemented through processors that work within this concept. Many functions are actually software subroutines or files stored in memory. This facilitates the implementation and adaptation to the specific needs of the network. The following is the list of functions that are part of the *5G NGC* (New Generation Core) cast. We will, however, group the functions together.

Here are the *Control Plane Functions* that are equivalent to those performed within the LTE/EPC architecture:

AMF (Access and Mobility Management Function), which is equivalent to the *MME of* the EPC, i.e., is responsible for connecting and managing user access, mobility, access authentication and authorization, and location service. Based on the service requested by the user, it selects the respective SMF to manage the session context.

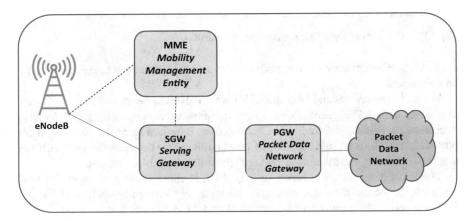

Fig. 7.15 CUPS (control and user plane separation) purely logical separation (inline)

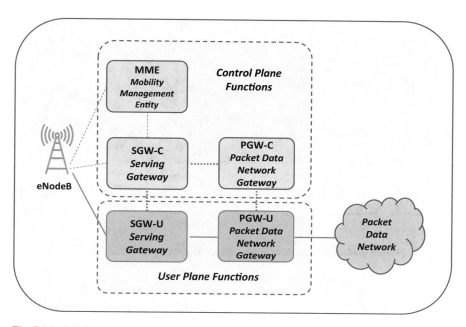

Fig. 7.16 CUPS (control and user plane separation) co-located separation

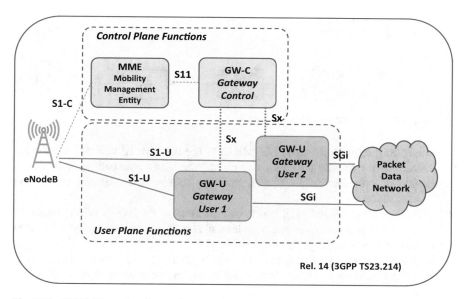

Fig. 7.17 CUPS (Control and user plane separation) remote separation

SMF (Session Management Function), which is equivalent in part to PGW and MME of the EPC, manages each session (negotiation of the establishment of a

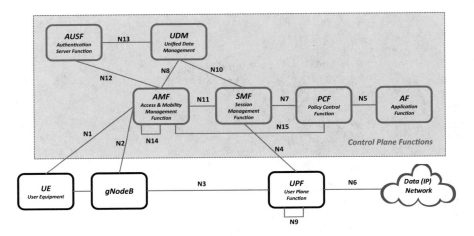

Fig. 7.18 5G non-roaming system architecture/reference point representation

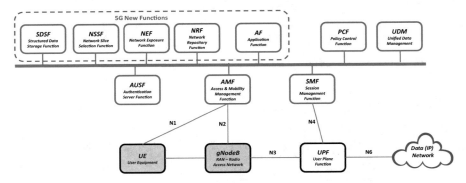

Fig. 7.19 5G-SBA (service based architecture)

connection) requested by the UE, allocates the network IP address, selects the User Plane for routing purposes, and directs the user's connection, controls the quality of communication (QoS), and the requirements demanded by the Control Plane.

PCF (Policy Control Function), which is equivalent to the EPC's PCRF, manages the network operation rules and policies and accesses information about the contracted service to manage the network operation.

UDM (Unified Data Management), which is equivalent to the EPC's HSS/AAA, manages user data and user identity, including the generation of authentication credentials and authorization for roaming.

AUSF (Authentication Server Function), which is equivalent in part to the EPC's HSS/AAA, is what allows the AMF to authenticate the user for network access.

The following Control Plan functions have no corresponding functions in the LTE/EPC:

SDSF (Structured Data Storage Network Function) is an auxiliary service used to store data in a structured way, which can be implemented, for example, by a SQL (Structured Query Language) database.

UDSF (Unstructured Data Storage Network Function) is another auxiliary service used to store data in an unstructured way and that can be implemented by a Key/Value Store (database), very interesting for OOP (Object Oriented Programming) programs.

NEF (Network Exposure Function), which allows the connection with service systems provided by third parties, performing the translation of external x internal information. This access can be implemented by an API (Application Programming Interface), which is nothing more than a set of routines and standards established by a software to allow the use of its features by other applications.

NRF (Network Repository Function) is a function designed to register and identify available Network Function (NF) services, enabling NFs to identify appropriate services within the Core, which can be implemented by a Discovery Service. A Discovery Service is a service that automatically identifies devices and services that can be provided by another device that is part of a computer network.

NSSF (Network Slice Selection Function) allows the selection of a "Network Slice," i.e., the slice among the available alternatives that best meets the customer's service needs. It is important to remind that 5G offers differentiated services and that network slicing allows to ensure better performance for the user's application, in addition to organizing network traffic and maintaining its efficiency.

Briefly, the first group of functions presented corresponds to a reorganization of already known LTE/EPC functions, while the second group concerns a list of new functions that 3GPP introduced with the objective of allowing the addition of new services provided by third parties and their integration with the 5G system.

Finally, we leave the context of the functions relating to the Network Control Plane to enter the realm of *User Plane Functions*.

UPF (User Plane Function), which corresponds to the S/PGW of the LTE/EPC network, whose function is to route all user content packets so that they travel between the internet and the RAN (Radio Access Network). Besides being responsible for forwarding messages, it also fulfills the network rules (Enforcement) and, by reporting the use of services aiming at their billing, guarantees the quality of the Service provided (QoS) and intercepts communications when authorized by law. There was also a concern to enable the construction of 5G networks for "cloud" processing within a Data Center designed to house the core of the cellular network.

As mentioned above, the joint operation of the two networks LTE and 5G is of fundamental importance, especially at the start of the operation of 5G networks, which will probably have to integrate with the existing LTE infrastructure. That said, we can begin to imagine how the installation and operation of *5G NSA* (Non-Stand Alone) and *5G SA* (Stand Alone) systems will happen.

Several alternatives of simultaneous or joint operation of the two networks are foreseen. In simultaneous operation, the two Cores remain operational, but need to communicate. To allow this configuration, it is necessary to adopt the architecture of 5G in Non-Roaming format, which was presented first. This denomination adopted by 3GPP is interesting because it gives the impression that it would not be possible to have roaming between the two networks. Obviously, this is not what happens, but for roaming to be possible, they need to be interoperable. Remember that in the Non-Roaming architecture communication between the functions is done through point-to-point connections and not through the bus, as is the case with the *SBA* architecture (Service Based Architecture), keeping the same concept of EPC. Certain functions common to both networks, already presented above, are unified, but each one performs its tasks respecting its own architecture. The user connects to the network independently in each of the systems and can achieve mobility (roaming), thanks to the interaction of the various modules and functions of the set, which allows interoperability between the two networks efficiently. This interoperability, which respects each of the architectures, is fundamental in networks using millimeter waves (mmW), where coverage is very restricted, and which will be able to use the existing LTE infrastructure to provide wider coverage.

Figure 7.20 shows how to provide interoperation between the two networks and how to use the functions common to both.

Note the appearance of a new connection, the N26, created to provide the interaction between the EPC's MME and the NG-5GC's AMF (Next-Generation 5G Core). This connection replaces the existing S10 interface in the EPC, which is used to connect two EPCs, as previously shown. This connection is critical to provide interoperability between the two Cores, but it is not essential. If it does not exist, it means that only one Core will be responsible for managing the network, as we will see below.

This architecture implies the fact that the two networks are operating simultaneously. But, as already pointed out, since the two technologies are originally similar, it makes perfect sense to evaluate how they can converge and become a single entity. 3GPP introduced six architecture options that the operator can use as a reference to redesign its network, in addition to the option of interoperation of the two generations, as just explained. The six options are shown in Fig. 7.21.

It is important to note that the alternatives can be grouped in two different ways: the Stand-Alone group (alternatives 1, 2, and 5) and the Non-Stand-Alone group (alternatives 3, 4, and 7). Each operator will have to seek the best alternative for his system. But for sure, this will be a dynamic process, which will evolve according to the needs and demands of the users. Thinking coldly about the efficient use of spectrum, it is not difficult to envision soon 5G, due to all its technological innovations, will operate in most of the frequency bands currently destined for other generations of cellular services. This situation would be ideal for any operator in terms of network operating costs. The major hurdle, besides the new investments required to upgrade networks, will be the migration of the UE, as User Equipment should be able to operate in this new exclusively 4G + 5G network.

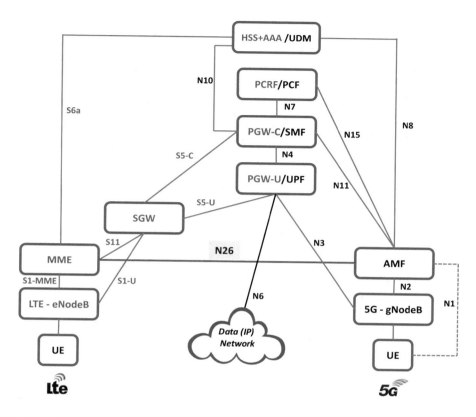

Fig. 7.20 Interoperation between LTE/EPC and 5G core

For the importance of this matter, let's evaluate each of the alternatives. Obviously, there is not much to say about *Option 1 – SA*, because it is simply a traditional LTE network.

Option 2-SA is basically the situation of an operator that does not own any legacy technology, which is called greenfield operations. In this case, it starts from scratch and the entire installation is done according to 5G standards. Since its deployment, the system can provide all expected services, including traditional services and the three fundamental types of new services – eMBB, mMTC, and URLLC.

Closing the Stand-Alone alternatives, we have *Option 5 – SA*, in which the LTE Base Station is connected to the 5G Core (5GC) through the NG-C and NG-U interfaces (Next Generation Control Plane and User Plane). In this option, the EPC is simply replaced by the 5GC. For this to be possible, the eNodeB software must be modified, upgraded, to be able to interoperate with 5GC in such a way as to become a ng-eNB (next generation – eNB). This will allow the provision of the differentiated services of Network Slicing, but the system will not be able to benefit from the advantages of the 5G NR (5G New Radio) air interface, such as multiple

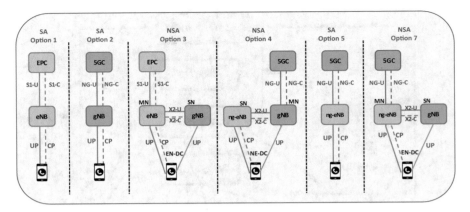

Fig. 7.21 Options envisaged for redesigning LTE and 5G networks

Numerology, already explained before and which provides greater spectral efficiency. The ng-eNB is therefore still an LTE Base Station. This is an alternative that is unlikely to be applied.

We will now evaluate the Non-Stand-Alone alternatives, which should occur in the so-called brownfield operations: situations in which an operator already uses LTE technology and needs to integrate its system with the new technological generation, i.e., intends to unify the two technologies, probably by integrating 5G to the LTE architecture and, subsequently, reversing the situation and integrating LTE to 5G.

Option 3 – NSA would most likely be the first stage of deployment of 5G gNB BSs, a situation where 5GC has not yet been deployed. For this to be possible, it is necessary that the eNB can work in *EN-DC* (E-UTRA-New Radio-Dual Connectivity) configuration, so that the eNB can communicate with the en-gNB through the known X2 interface (X2-C/X2-U), the same used to interconnect two eNBs. In this case, the eNB becomes the Master Node (MN), while the en-gNB acts as the Secondary Node (SN). The two Base Stations, eNB and en-gNB, can provide the User Plane protocols, while the Control Plane function is provided by the eNB for communication with the UE. Option 3-NSA does not support Network Slicing and consequently, URLLC cannot be implemented. But this option extends the traffic capacity.

The nomenclatures *en-gNB*, cited above, and *eg-eNB*, which will be cited below, appear to indicate that these Base Radio Stations belong to two different generations and have the ability to communicate with each other and with the Core of the two networks, EPC or 5GC.

Option 4-NSA is almost a mirror image of Option 3-NSA, i.e., the positions are reversed. It tends to be the option to be deployed in the future, when the 5GC Core is sedimented. At that time, the 5G network should already be in operation, and the LTE Radio Base Station (ng-eNB) should be connected to the 5G Radio Base Station (gNB) in a secondary character, Secondary Node (SN). The 5G gNB acts as the Master Node (MN). The ng-eNB shall be modified to allow the use of the Xn communication channel (Xn-C/Xn-U), which will allow communication with the gNB. In this situation, the connectivity is *EN-DC* (E-UTRA New Radio-Dual Connectivity). In this case, Control Plane functions are provided by the gNB, while User Plane functions can be supplied by either the ng-eNB or the gNB.

Finally, in *Option 7-NSA*, it is the ng-eNB that communicates with the Core 5GC and acts as the Master Node, while the gNB acts as the Secondary Node. This is the so-called *NGEN-DC* (NG-RAN E-UTRA-NR Dual Connectivity) connectivity. The Control Plane connection is maintained by ng-eNB, while User Plane functions can be provided by any Base Station.

Chapter 8
The Protocols Used: The Language Spoken by the LTE System

So far, we covered, basically, issues related to hardware responsible for the operation of transmission and reception of electromagnetic signals, including antennas. We have shown how the architecture of the various generations of mobile phones and described the main components that are part of these systems. We showed each of the elements of the structure statically without going into detail about the dynamic part, or how the parts "talk" or exchange information with each other to allow communication to occur through the cellular network efficiently and with quality.

The multiple dialogues that occur during the transmission and reception of messages is an extensive and rather tedious subject, since it involves a huge number of details whose approach would not make sense at the level we have proposed to present in this book. We will, however, analyze the fundamental points of the subject in order to allow the reader to have an idea of the dynamics of the main dialogues that occur between the main participants of the network. As the subject is fundamentally related to an IP access network, we have organized in Chap. 13 a description of how a network works within its more generic concept. The LTE or 5G network is also based on the same principles as the IP network, but its functionalities are quite peculiar, since these networks operate according to their own model, which we will describe later.

Before going into the details of the network structure, let's describe how traffic flows, when a device is turned on and seeks to be served by the network, and then talk about how it is handled at each node by the respective elements, within levels that perform very specific functions.

The most crucial concept to understand how data traffic works in an LTE network, and therefore also in a 5G network, is to understand the meaning of the term *BEARER*. We already had the opportunity to explain what a Bearer is before, and we associate its image to a tunnel that allows communication to flow in a continuous and express way.

In the LTE network, the Bearer is used to establish a communication channel between the user equipment (UE) and the packet network passing through some

© The Author(s), under exclusive license to Springer Nature Switzerland AG 2023
J. L. Frauendorf, É. Almeida de Souza, *The Architectural and Technological
Revolution of 5G*, https://doi.org/10.1007/978-3-031-10650-7_8

network elements but is also associated with the quality (QoS) requirements associated with the content that will flow in the network.

As said, the most intuitive way to understand a bearer is to view it as just such a tunnel, or a direct connection that will link the user equipment (UE) to the Packet Data Network (PDN), i.e., an almost direct way to access an external IP network. Several tunnels are established in a concatenated manner, as shown in Fig. 8.1. An EPS Bearer is always associated with a Quality of Service (QoS) level, and a single user may be connected to multiple Bearers simultaneously. The reason is simple: the user may be accessing multiple services at the same time. They may be talking on the phone, using VoIP service, accessing a private network, a public network, listening to music, or downloading a video or a file. Each application demands different requirements and, consequently, each service is linked to its own QoS. Although there is no limit to the number of Bearer classes, nine classes of the so-called *QCI* (Quality of Service Class Identifier) were created. This classification considers the type of service, the sending priority, and limiting parameters, such as Packet Loss, due to transmission errors and Delay (Packet Delay).

Each service is associated with a QCI, and each QCI is linked to a certain level of transmission preference programmed in the network. The system is equipped with a mechanism called *ARP* (Allocation and Retention Priority), which defines the priority, allocation, and maintenance of Bearers in the network. This mechanism is very important to determine the establishment or rejection of a new Bearer when the network is congested. Bearers can be associated with two basic priority categories: Minimum Guaranteed Bit Rate *(GBR)* communications, for applications such as VoIP, video chat, or real-time games, for example, and *Non-GBR* Bearers, for which there is no guaranteed bit rate. Table 8.1 shows the different Quality of Service Class Identifier classes:

Having explained the concept of Bearer, let's understand how a user "connects" on the network (*Attach Procedure*) and how to get the service in acceptable quality standards.

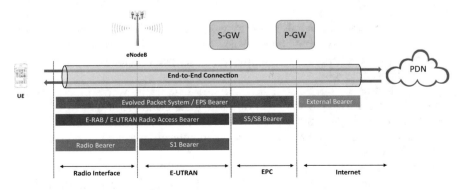

Fig. 8.1 Bearer communication channel concept

Table 8.1 Different QCI (Quality of Service Identifier) classes

QCI	Bearer type	Priority	Delay (ms)	Packet loss	Example
1	GBR	2	100	10^{-2}	VoIP
2		4	150	10^{-3}	Video
3		3	50		Online games (real time)
4		5	300	10^{-6}	Video streaming
5	Non-GBR	1	100		IMS signaling
6		6	300		Video, TCP services (email, chat, FTP, etc.)
7		7	100	10^{-3}	Voice, interactive games
8		8	300	10^{-6}	Video, TCP services
9		9			(email, chat, FTP, etc.)

The Cellular Network needs to identify and exchange information with the subscribers within its coverage area. For doing so some identifiers are used, among which we highlight:

- *Network Identity* – The *Network ID/PLMN* (Public Land Mobile Network) identifies the network internationally and is comprised of the *MCC* or Mobile Country Code and the *MNC* or Mobile Network Code.
- *Tracking Area ID/TAC* (Tracking Area Code) – Identifies the tracking area formed by a group of cells.
- *Cell ID* – Identifies the cell.

Periodically, the Base Station (eNB) provides information within its coverage area so that it can locate subscribers. In addition to identifiers, the network also provides information about the radio interface and cell capabilities and competencies. The first step that the device performs when powered on is to "listen", to receive the information provided by the system within its geographical perimeter. To this end, the eNB constantly transmits (broadcast) the *SI* (System Information), which are nothing more than messages containing information about the system that are transmitted to all user devices. This enables the device to identify the network to which it intends to connect. The first step is for the device to synchronize with the system. This information is provided in the form of information blocks, the so-called *MIB* (Master Information Block) and *SIB* (System Information Block), which allow the UE to find and synchronize with the network.

Once the user equipment is synchronized and knows the characteristics of the cell from which it wants to obtain services, it is now able to initiate a dialogue by sending the *RAP* (Random Access Procedure) message. There are two formats of RAP. One is the *Contention-Based RAP*, a process that occurs when several UEs try to communicate simultaneously generating collisions that are overcome through an initial dialogue between the eNB and each of the UEs, organizing the contacts in a staggered manner to circumvent the problem. The other procedure is the *Non-Contention-Based RAP*, when the cell itself promotes the handover between nearby cells.

Once the initial conflict is overcome, the dialogue between the UE and the eNB, the *RRC* (Radio Resource Control Contention Set-up) is initiated as shown in Fig. 8.2.

Once the dialogue between the UE and the eNB is established, it is possible to establish contact with the rest of the network, that is, the Core EPC. In short, initially, we have an exchange and signaling phase centralized by the MME (Mobility Management Entity) to allow, or not, the user connection in the cellular network. In addition, the necessary resource reservation is made so that the UE can be connected to the packet/internet network. Then, at the end of the signaling phase, in the first stage, a "default" Bearer is established to allow the user to exchange information directly with the packet network. New Bearers appropriate to the services or applications requested by the user (UE) can be requested as needed, as shown in Fig. 8.3. The connection is always made with the incoming port of the Core EPC user plane, i.e., the *S-GW* (Serving Gateway), and the outgoing port, the *P-GW* (Packet Data Network Gateway).

The message exchange is quite extensive. Follow the sequence of messages in Fig. 8.3 to understand the description more clearly. The sequence of the dialogue is as follows:

Once the *RRC* is complete, the UE sends the *Attach* request (network connection) to the eNodeB, and the eNodeB passes the request to the MME, which is

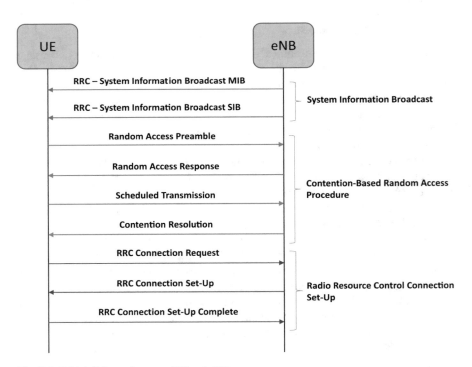

Fig. 8.2 Initial dialogue between UE and eNB

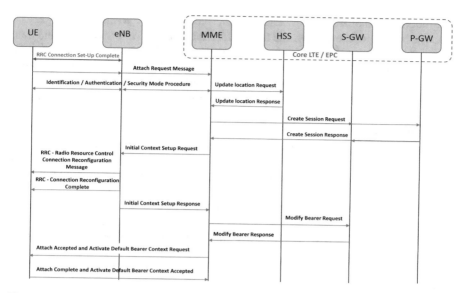

Fig. 8.3 Attach procedure dialogue (Core attachment)

responsible for the dialogue control (signaling) and activity management. The MME will use the information stored in the *HSS (*Home Subscriber Server) to allow, or not, the subscriber registration on the network. In possession of the necessary information, the MME will exchange information with the user (UE) to perform the authentication process. Once the subscriber has been successfully authenticated and the security criteria established, the MME sends an *Update Location Request* message to the HSS requesting the update of the UE information and receives an *Update Location Response* message indicating that the data has been updated. Then, the MME sends a *Create Session Request* message to the S-GW, requesting that a session be created for the user's connection. A session is nothing more than a communication channel allocated to a particular activity for a certain period. The S-GW selects a P-GW, which in turn begins the negotiation with the packet network, being this the moment when an IP address is assigned to the UE. The *Create Session Response* message is sent to the MME indicating that the session is created. The MME initiates a new dialogue with eNodeB by sending the *Initial Context Setup Request* message with the objective of establishing the context, that is, to determine the characteristics of the communication, the resources involved, and the capabilities required. At this point, the UE and the eNodeB establish the security parameters required for data encryption so that the messages exchanged via radio are encrypted and have their integrity protected.

   The eNB sends information regarding the reconfiguration of the radio resources that have been allocated to the UE, and the UE acknowledges receipt of the information. With this the communication setup is complete, then the *Initial Context Setup Response* message is sent. Next, the MME sends a message to the S-GW requesting the modification of the Bearer state with the *Modify Bearer Request* message. The S-GW responds to the request and the MME with *Modify Bearer Response* and then communicates with the UE, informing that the connection to the network is completed by sending the *Attach Accepted and Activate Default Bearer Context Request* message. The UE then sends the *Attach Accepted and Activate Default Bearer Context Accepted* message. With that, this step is complete, and the user can now use the packet/internet network services.

There are several other similar procedures that refer to the handover dialogues, network disconnection, user terminal disconnection, and dialogue of answering a phone call. It would be pointless to detail each of the procedures here. The intention was to use the Attach Procedure only to exemplify how a user connection dialogue to the network occurs.

Understanding how the dialogue happens, it's time to talk about the network hierarchy, i.e., who does what, and who is responsible for receiving and processing the data received. We have already explained that LTE, as well as 5G, work within the concept of a network and that there are several levels of dialogue between the different entities that make up the network. These levels are the various layers, like those used in TCP/IP and OSI networks.

Regardless of the dialogue between the elements that make up the network, it is interesting to quickly recall what the functions of its main components are, as shown in Fig. 8.4.

The protocol that runs between the user equipment (UE) and the eNodeBs is called Access Stratum Protocol (*AS*). The eNodeB is fundamentally responsible for controlling the radio resources in the LTE architecture. One of its main functions is the *RRM* (Radio Resource Management), which controls all resources related to the

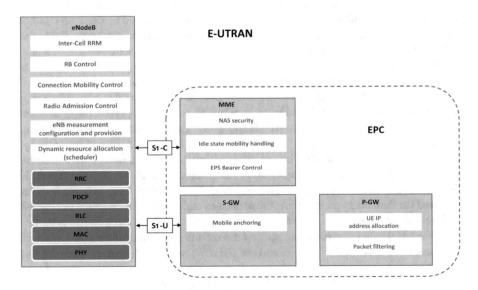

Fig. 8.4 Main functions of eNodeB and the EPC Core

air interface, including the coordination of radio resources between different cells (Inter-Cell RRM), which allows better use of the available spectrum (Space Division Multiple Access), already explained before. This includes the reuse of frequencies avoiding interference between cells (ICIC – Inter-Cell Interference Coordination), Bearer Control, Admission Control in handover, and control of *RB* (Resource Block), which, as we have seen, is the smallest unit of transmission resources.

Another important control is the scheduling of uplink and downlink messages. A special dynamic resource allocation control algorithm (Dynamic Resource Allocation) controls the priority of messages from different users and, consequently, manages the RB insertion in the transmission flow.

There are other functions also performed by eNodeB, besides controlling radio resources and user mobility. Header Compression is one of them. It takes care of reducing the size of the transmitted packets as much as possible by reducing redundant header data, especially small data packets in which the contents, as in VoIP, are greatly reduced. Another important function is the encryption of the messages to ensure security at the air interface.

Perhaps one of the most important functions performed by eNodeBs is the dialogue between them, performed by the X2 connection interface. Thanks to this dialogue, it is possible to reduce the latency time of information from a user moving within the mobile network's operating area. The user information (context) and buffered data are transmitted from one eNodeB to another without passing through the Core.

The eNodeBs also provide information about measurements, provisioning, and configurations.

Moving on now to better characterize the functions of the Core, we recall that the *S-GW* (Serving Gateway) is its gateway and that it also acts as an "anchor" point every time a user moving around the network is disconnected from an eNodeB to connect to another eNodeB.

The *MME (*Mobility Management Entity) is the unit that processes all the UE signaling in the Core EPC. The communication protocol used between the UE and the MME is the NAS (Non-Access Stratum). The NAS is so named to differentiate it from the already mentioned AS (Access Stratum), the protocol responsible for transporting information that passes through the air interface between the UE and the eNodeB. The MME manages all the functions related to mobility and the state of the user's equipment within its coverage area in the network. Another important function is the management of Bearers. Once the UE is registered in the network, that is, once it has fulfilled all the Attach Procedure explained above, the UE can be active or in a standby state. When on hold, the UE enters the ECM-Idle (EPS Connection Management Idle) state, and the MME maintains the Context and all information regarding the established Bearers. The UE continuously updates its position on the network every time it moves out of its current *TA* (Tracking Area) in a procedure called *Tracking Area Update*. The MME is responsible for tracking the user's location while the UE is in ECM-Idle. When a new message is destined for the user (idle), the MME sends a paging message to all eNodeBs within its management area (Tracking Area), while the latter (eNodeBs) communicate with the UE over the radio interface. Upon receiving the call, the UE performs the *Service Request Procedure,* when its status changes to ECM Connected and the Bearers are reestablished. The MME is the one who commands this reestablishment and updates the UE's Context on the eNodeB. There is a special, differentiated procedure for the case of an emergency call, *Emergency Attach.*

After completing the review of the most prominent functions performed by the main blocks, let's talk about the *P-GW* (Packet Data Network Gateway). This is the gateway from the access network to the packet network. It is responsible for allocating an IP address and filtering the packets received so that they are classified according to the needs of quality of service (QoS) and according to the characteristics of each of the Bearers.

We can now enter the detailing of the layers (Layers) used throughout the communication of the LTE system. There are three: L1, L2, and L3.

- L1 – *PHY* (Physical Layer)
- L2 – It is subdivided into three layers:

  - *MAC* (Medium Access Control)
  - *RLC* (Radio Link Control)
  - *PDCP* (Packet Data Convergence Protocol)

- L3 – *RRC* (Radio Resource Control)

There is a supplementary layer used exclusively by the Control Plane. This supplementary layer is called *NAS* (Non-Access Stratum), as already mentioned, and directly connects the UE to the Core EPC, specifically to the MME, being transparent to eNodeB. It is important to highlight something already explained previously: the LTE separates the control data (CP – Control Plane) from the content data (UP – User Plane). Figure 8.5 shows how the communication regarding Control Plane and User Plane is done.

So, let's learn more about each of the components of these stacks.

NAS (Non-Access Stratum) Unique layer of the Control Plane, it manages all signaling between the UE and the MME in the Core EPC and is used to establish, maintain, and terminate the UE session, supporting user mobility on the network (active) and the UE in idle state (Idle Mode).

RRC (Radio Resource Control) Manages the establishment, maintenance, and interruption of RRC connections between the UE and the eNodeB, broadcasts system information, promotes UE Paging, participates in the handover between cells, and selects which of them should provide the service based on information about the quality of the UE connection with the various cells, participates in the integrity protection and encryption of RRC messages, and manages the RRC security keys (different from the user plan keys), controls the Bearers and the maintenance of the QoS requirements of the different active radio Bearers, carries out the UE measurement report, and signal quality control.

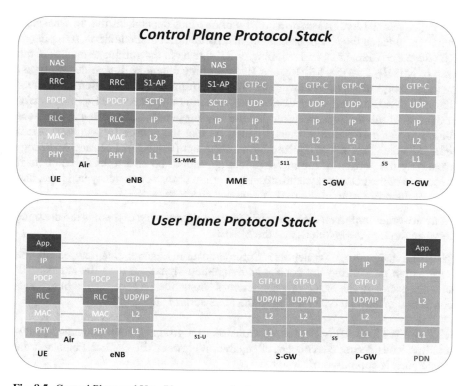

Fig. 8.5 Control Plane and User Plane communication

PDCP (Packet Data Convergence Protocol) Manages message encryption, compression/decompression of headers, and the sequencing of PDUs (Packet Data Units), taking care that they are in the right sequence, since they are not necessarily transmitted in the correct logical sequence.

RLC (Radio Link Control) Manages the segmentation of data packets (PDUs), taking care that resending occurs when a PDU arrives corrupted. This is done through the Automatic Repeat Request (*ARQ*) mechanism. LTE takes care of the sequencing of the *SDUs* (Service Data Unit) of several users, ordering its transmission in order to make the air interface as efficient as possible. This procedure compromises latency and, for this very reason, was reformulated in the case of 5G.

MAC (Medium Access Control) Manages the multiplexing of data from the various logical channels inserted or extracted in the transport blocks to be forwarded to the physical layer, detecting transmission errors and requesting retransmission through *HARQ* (Hybrid Automatic Repeat Request), mapping between Logical Channels and Transport Channels, as we will see later, manage priorities according to the Dynamic Scheduling algorithm and promotes message "padding" with SDUs of irrelevant content when there is nothing to transmit.

PHY (Physical Layer) Performs all functions that depend on the air interface, which includes setting the transmission rate, modulation/demodulation, mapping of RE (Resource Elements), and mapping of RF signals to the various antennas, in the case of MIMO, Massive MIMO, and Beamforming antennas. It also detects transmission errors and reports to the upper layers and promotes the insertion and extraction of FECs (Forward Error Correction), which is a method used to allow error correction at the receiver ensuring that the message sent can be recovered without failure at the destination, because it may be that the source (transmitter) sends redundant data that are recognized by the recipient (receiver). The receiver recognizes only the portion of the message that appears to be error-free. In addition, it takes care of the network synchronization, both in terms of frequency and time, and does all the processing of the radio frequency channels.

There are other important dialogues within the network. One of them is the dialogue between the UE, the eNodeB, and the MME:

S1-MME (Signaling Transport) Communication channel that carries all the information concerning the control and signaling signals. It uses the standard IP network protocol, the *SCTP* (Stream Control Transmission Protocol), whose function is to transport all data relating to the NAS control signaling.

GTP (GPRS Tunneling Protocol) Consists of a group of communication protocols that, considering the concept of separation between the Control Plane and the User Plane, is composed of:

- *GTP-C* (Control), a protocol used by the Control Plane that ultimately enables network access by activating a PDP (Packet Data Protocol Context Activation) session, deactivating it, adjusting QoS-related parameters, and updating the session when user communication is being transferred from another EPC.
- *GTP-U* (User), protocol used in the User Plane. GTP-U tunnels are used to transport encapsulated Transaction Protocol Data Unit (TPDUs) and signaling messages between the two endpoints of the GTP-U tunnel, i.e., Bearer data packets are transmitted peer-to-peer through the tunnel formed between eNodeB, S-GW, and P-GW. The Tunnel Endpoint ID (TEID) is present in the GTP header to indicate to which tunnel a given TPDU belongs.

As the various specific layers perform their functions, the network is becoming like a conventional IP network, in which stands out the *SCTP* layer (Stream Control Transmission Protocol), which is a conventional signaling protocol used in the IP architecture. This protocol is important because it allows the transmission of all signaling exchanged between eNodeB and MME in a safe and efficient way, since these signaling packets travel over the S1-MME interface (backhaul), which is based on the IP network. SCTP is also part of the Diameter protocol stack, used in the interface between the MME and the HSS to promote authentication, authorization, and service accounting. Within the Core EPC, for connections between MME, S-GW, and P-GW, the protocol used is *UDP* (User Datagram Protocol), also present in IP networks, and has the advantage of being fast, simple, efficient, and with little transmission overhead when compared to *TCP* (Transmission Control Protocol). Although efficient, it does not allow the retransmission of packets that may have been lost.

Before closing the chapter on LTE, it is worth showing what are the LTE Communication Channels, their classification, and how they allow communication between all layers of the network. Figs. 8.6 and 8.7 facilitate the understanding.

There are basically three types of communication channels:

Logical Channels Defines *what type of* information is being transmitted over the air. These channels are defined according to the data transfer services offered by the MAC Layer. Messages containing data (User Plane) and signaling (Control Plane) are carried on the logical channels between the MAC and RLC layers. They are distinguished by the *type of information* they carry and can be classified into two categories: *Control Channels*, for Control Plane messages, and *Traffic Channels*, for User Plane messages. Control Channels can be common to all users, broadcast (Point-to-Multipoint) or dedicated (Point-to-Point). Table 8.2 shows what these messages are.

Transport Channels Channels that define *how data is transferred* over the Physical Layer and *what are its characteristics*. Both data and signaling are transported between the Physical and MAC Layers. The Transport Channels are distinguished by how the processor handles these channels. Table 8.3 shows what these messages are.

Physical Channels Channels that carry the information originating from the upper layers and which are distinguished by indicating *where* the information is routed, *contain data messages and signaling*. They are divided into two subgroups:

- *Physical Data Channel*, channels that are distinguished by how the Physical Channel processor handles them and *how data is allocated to* symbols and subcarriers when mapped during the OFDMA modulation process. Table 8.4 shows what these messages are.
- *Physical Control Channels* – The processor of the Transport Channel of this layer generates the control information necessary to support the operations of the Physical Layer. This information is sent to the processor that controls that channel. The information travels to the Transport Channel processor at the receiver but is completely invisible to the higher layers. Similarly, the Physical Channel processor creates physical signals, which support the lower layer aspects of the system. The list of these channels is shown in Table 8.5.

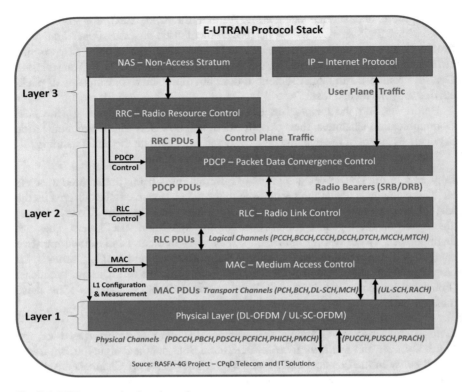

Fig. 8.6 LTE communication channels structure

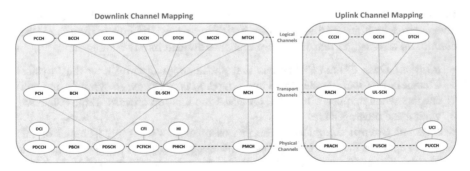

Fig. 8.7 LTE communication channels (Downlink and Uplink)

The processor on this channel uses a series of control information to support the operation of the lower levels of the Physical Layer. These are listed in Table 8.6.

There are still the *Physical Signals*, which are signals issued by the eNodeB Base Station and that help the mobile user units (UE) at the time they are turned on, because these signals are of fundamental importance to promote the synchronization of communications. They are *PSS* (Primary Synchronization Signal) and *SSS* (Secondary Synchronization Signal).

Table 8.2 Logical control channels/control plane and traffic channels/user plane

Channel name	Abbreviation	Control channel	Traffic channel
Broadcast control channel	BCCH	X	
Paging control channel	PCCH	X	
Common control channel	CCCH	X	
Dedicated control channel	DCCH	X	
Multicast control channel	MCCH	X	
Dedicated traffic channel	DTCH		X
Multicast traffic channel	MTCH		X

Table 8.3 Transport channels

Channel name	Abbreviation	Downlink	Uplink
Broadcast channel	BCH	X	
Downlink shared channel	DL-SCH	X	
Paging channel	PCH	X	
Multicast channel	MCH	X	
Uplink shared channel	UL-SCH		X
Random access channel	RACH		X

Table 8.4 Physical data channels

Channel name	Abbreviation	Downlink	Uplink
Physical downlink shared channel	PDSCH	X	
Physical broadcast channel	PBCH	X	
Physical multicast channel	PMCH	X	
Physical uplink shared channel	PUSCH		X
Physical random access channel	PRACH		X

Table 8.5 Physical control channels

Channel name	Abbreviation	Downlink	Uplink
Physical control format indicator channel	PCFICH	X	
Physical hybrid ARQ indicator channel	PHICH	X	
Physical downlink control channel	PDCCH	X	
Physical uplink control channel	PUCCH		X

Table 8.6 Additional physical transport/control channels

Channel name	Abbreviation	Downlink	Uplink
Downlink control information	DCI	X	
Control format indicator	CFI	X	
Hybrid ARQ indicator	HI	X	
Uplink control information	UCI		X

Chapter 9
The Protocols Used: The Language Spoken by the 5G System: Differences Between 4G and 5G Architectures

Typically, in LTE architecture, the eNodeB is composed of the radio part (RRU), located near the antenna, and the BBU, which is near the base of the tower, and the systems are proprietary both in terms of hardware and software.

The EPC (LTE Core) is generally monolithic, centralized, and far away from the eNodeBs. This architecture makes it difficult to reduce latency and overloads the communication paths of both the fronthaul and the backhaul.

With the advent of 5G, the Core has been divided into functions, with each function being able to be processed independently of each other. Perhaps most importantly, they can be processed on "off the shelf" hardware and servers (COTS – Common Off The Shelf hardware).

With the growth of IoT applications and new services, which require lower latency, the need for decentralization of Data Center functions has been presented, in a way that some computational functions related to processing and storage of service data can be moved to mini–Data Centers located in positions closer to the network edge shortening the path to be traveled by the data increasing the speed of communications and reducing latency.

In 4G there was a concern with quality of service (QoS) and differentiation of the quality of service provided according to the subscriber's classification, but in 5G the level of commitment to the quality of communication offered for a given type of service is much higher. In addition, the network can be customized, according to the services it intends to provide, within the concept of slices, which is something new in 5G. Recall, in 5G, the operator may have to serve several types of services with characteristics quite different from each other. It can provide highly reliable services with extremely low latency, URLLC – Ultra-Reliable Low Latency Communications, or meet the demand of a large number of IoT (Internet of Things) device connections that do not have such strict latency or speed requirements, being called mMTC – massive Machine Type Communications, or still, meet the demand of services that require large volumes of data, the eMBB – enhanced Mobile

© The Author(s), under exclusive license to Springer Nature Switzerland AG 2023   123
J. L. Frauendorf, É. Almeida de Souza, *The Architectural and Technological Revolution of 5G*, https://doi.org/10.1007/978-3-031-10650-7_9

Broadband. A new concept of service provision emerges within a new strategy to obtain profits based on the classes of services provided.

In a specific chapter, we showed how the architecture of the various generations of cellular services has evolved and made a point of showing the extent to which 4G and 5G can share the same Core, although this should not be the operator's ultimate goal when deploying a 5G network. We have seen that it would be possible for 4G and 5G networks to evolve until they could be fully integrated. We will now show the fundamental differences involving the management of the networks, which, although they arose from the same DNA, differ fundamentally in their respective implementations.

The 5G New Radio Core (NR) is designed not only to manage the access and mobility of 5G mobile phone subscribers, but also to enable the connection of new services, as well as to ensure higher levels of communication quality and to control packets in the network through special functions such as Network Slicing. The high specificity of its functions also stands out. Following the line already adopted by 4G, 5G adopts CUPS (Control and User Plane Separation), that is, the Control Plane functions are completely separated from those of the User Plane, which ensures system agility.

As Figs. 9.1 and 9.2 show, with respect to subscriber mobility management and carrier formation only, the protocol "stacks" of 5G are like those of LTE, but with an important addition.

In Control Plane, the LTE S1-AP interface corresponds, in 5G, to *NG-AP* (New Generation Application Protocol). In 5G, the User Plane gains an additional layer, *SDAP – Service Data Adaptation Protocol*, which was created to manage Quality of

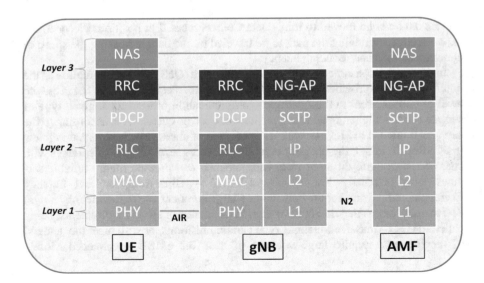

Fig. 9.1 5G Control Plane Protocol Stack

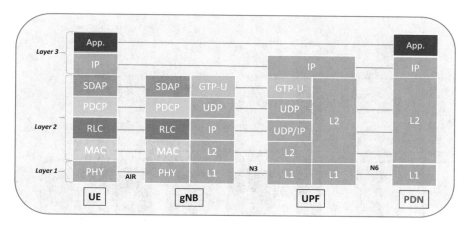

Fig. 9.2 5G User Plane Protocol Stack

Service (QoS) more precisely and is essential to ensure that the new applications proposed by 5G can be offered.

In Fig. 9.3, a summary of the role of each layer in both the Control Plane (CP) and User Plane (UP) can be found.

Figures 9.4 and 9.5, on the other hand, show us how QoS control is done in 4G and 5G.

Note that in 4G, there is a concern to identify the type of traffic that is traveling over the network and define a particular tunnel (Bearer) so that this traffic has a higher or lower priority in the sequencing of sending data packets. Note also that eNodeB does not participate in this process, which is managed exclusively by the UE and the PGW (Packet Gateway). In 5G, in addition to the criteria used by 4G, there is a second instance that is exactly the SDAP layer that is responsible for mapping the flow directly on the RBs (Resource Blocks). In this process gNodeB also participates. With this, it is possible to define traffic control parameters in the network as shown in the example in Fig. 9.6.

5G Quality of Service (QoS) model is based on QoS Flows. On the air interface, which uses AS (Access Stratum) protocol, the UE and eNodeB map QoS flows to Data Radio Bearers (DRBs). At the Non-Access Stratum (NAS) connection level, existing packet filters in UE and 5GC map UL and DL packets respectively to different QoS flows according to QoS rules. To each individual QoS flow is assigned to a unique identifier called the QoS Flow Identifier (QFI) and all packets tagged with the same QFI have the same forwarding. The standardized 5QI (5G QoS Identifier) values and their QoS characteristics specify basically three types of streams: Guaranteed Bit Rate (GBR) QoS Flows, Non-GBR QoS Flows, and Delay-Critical GBR. Every QoS flow has a QoS profile that includes QoS parameters and QoS characteristics. Applicable parameters depend on GBR or non-GBR flow type. QoS characteristics are standardized or dynamically configured.

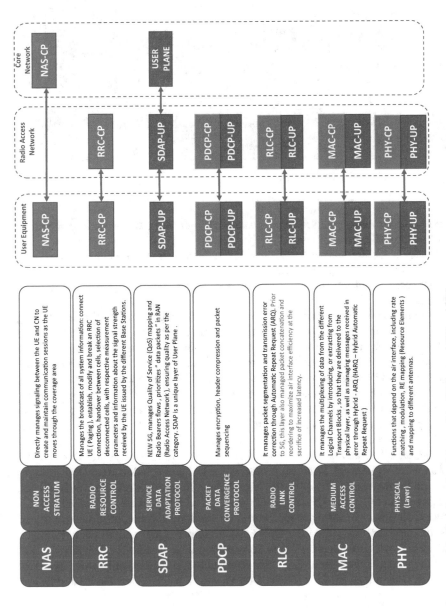

Fig. 9.3 Summary of the functions of each layer of the 5G Protocol (CP and UP)

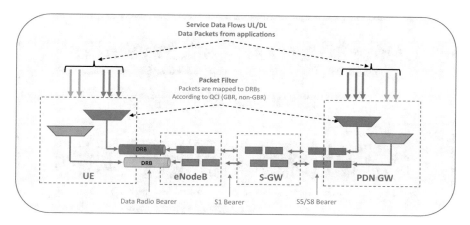

Fig. 9.4 QoS Model for 4G

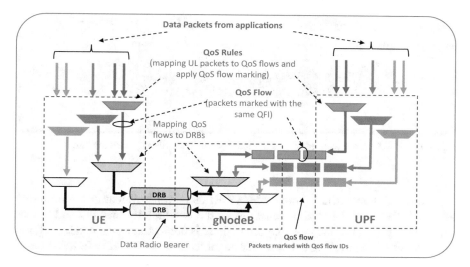

Fig. 9.5 QoS Model for 5G

In this new context, the granularity of control is much finer. It is possible to control the Quality of Service (QoS) of certain "logical slices" of the network, making it possible to treat the various services offered differently according to their classification. For example, slice 1 is reserved for packets classified as eMBB, slice 2 for packets whose services correspond to URLLC, and so on.

Network slicing can occur so that another objective can also be achieved, like sharing the same network among several operators in a totally isolated and secure way.

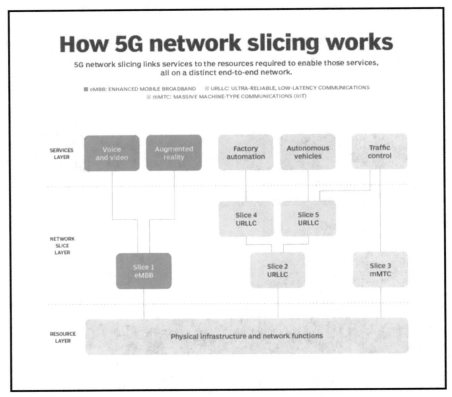

Fig. 9.6 5G QoS according to the Logical Slice. Source: Terry Slattery for Tech Target

Before going into detail about the 5G CORE, it is worth evaluating the similarity and equivalence of the various functions of the two architectures 4G and 5G, as shown in Fig. 9.7. But the most important thing is not to analyze in detail the functions, which are equivalent, but how they are processed. In LTE each function block connects to the other by means of an interface. This concept was kept in the *5G Non-Roaming architecture*, already shown previously only to allow the sharing of the two networks, especially with the EPC in the introductory phase of this new technology. But the great evolution came with the *5G-SA* (5G Stand Alone), which is implemented with the *SBA architecture (Service-Based Architecture)*, shown below. It is important to emphasize that this model is unique to 5G, and the EPC functions are shown only to illustrate the equivalences.

The LTE/EPC architecture typically works with few nodes (MME, PGW, SGW, HSS, PCRF, and AF), and each node is responsible for performing a specific function. Remember that each node is interconnected to the others through a specific interface. It is important to highlight that in 5G, *SBA* (Service-Based Architecture), is the only alternative that allows innovations. The Core has more than two dozen

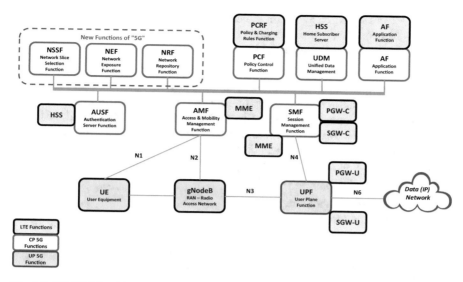

Fig. 9.7   5G CORE architecture and equivalence with LTE

network functions. Communication is done through a bus where all information from the Control Plane is transmitted. The main functions, shown in the figure above, are divided into two categories.

The first category gathers the *functions that find equivalence in the LTE architecture*. They are:

- *AMF* – Access and Mobility Management Function (≅ MME)
- *SMF* – Session Management Function (≅ MME + PGW-C + SGW-C)
- *UPF* – User Plane Function (≅ PGW-U + SGW-U)
- *AUSF* – Authentication Server Function (≅ HSS)
- *UDM* – Unified Data Management (≅ HSS)
- *PCF* – Policy and Control Function (≅ PCRF)
- *AF* – Application Function (≅ AF)

In a quick description of these functions, we can say that the AMF manages mobility, signaling, and subscriber access control; the SMF is responsible for session management, the UPF for routing and packet forwarding, the AUSF for the execution of the UE authentication processes, the UDM manages all user data, the PCF provides standards and traffic control policy, and the AF can be used as an auxiliary of dynamic policies and/or billing control. There is also a model that aggregates three functions into one, it is called *UDC* (Unified Data Convergence) which is the junction of the *UDM + AUSF + PCF* functions.

Three new *functions have been added* to support the new services proposed by 5G. They are:

- *NSSF* – Network Slice Selection Function
- *NEF* – Network Exposure Function
- *NRF* – Network Repository Function

The NSSF is responsible for the Network Slicing control functions and basically selects the network slice instance (NSI) and defines the AMF that will serve the UE. The NEF handles the masking of sensitive network and user information according to the adopted network policies. The NRF, on the other hand, provides registration and discovery of NF services (network functions), allowing NFs to identify appropriate services to each other.

A new category of *functions* was *created*. They are optional features that can be used by operators when needed, they are:

- *SDS* – Structured Data Storage
- *UDSF* – Unstructured Data Storage
- *BSF* – Binding Support Function (Broad Forward Support Function)
- *CHF* – Charging Function
- *NWDAF* – Network Data Analytic Function
- *NSSMF* – Network Slice Subnet Management Function
- *NSMF* – Network Slice Management Function
- *UCMF* – UE Radio Capability Management Function
- *UDR* – Unified Data Repository

Some new functions are intended to support interfacing with access networks that do not follow 3GPP (Non-3GPP Networks), which are:

- *N3IWF* – Non-3GPP Interworking Function
- *W-AGF* – Wireline Access Gateway Function
- *5G-EIR* – Equipment Identify Register

Finally, some functions have been provided to increase the security of the network:

- *TNGF* – Trusted Non-3GPP Gateway Function
- *TWIF* – Trusted WLAN Interworking Function
- *SEPP* – Security Edge Protection Proxy
- *SCP* – Service Communication Proxy

Again emphasizing, the adoption of these new functions can only occur if the SBA model is adopted. 5G makes extensive use of software resources and virtualization, for this reason, we will talk a little about this subject:

The *SDN* (Software Defined Network) architecture, which as its name expresses, is a network model fully defined by software functions and is based on network virtualization techniques, or rather *NFV* (Network Function Virtualization), which we will address below. First, however, it is important to note that many of these functions can be aggregated according to their intended purpose. Figure 9.8 shows how this mapping, according to a particular purpose, can occur.

There is a very special reason for these functions to be defined and standardized in the network according to their applications. The 5G architecture is identified by the suggestive name of SBA (Service-Based Architecture) and presents a revolutionary infrastructure designed specially to serve services with their own characteristics, besides adding a great advantage, that of being totally open, independent, and allowing the operator to choose the suppliers that best meet their needs. Therefore, manufacturers specialized in a certain group of functions contribute with a portion of the total solution, which added to the other functions developed by other suppliers, make up an optimized set that will meet the operator's needs. It is not difficult to conceive that those networks can be built from the expertise of two, three, four, or more solution providers, each considered best-of-breed. Figure 9.9 illustrates this possibility.

In the language of computing, the communication between two applications is done by *API* (Application Programming Interfaces). An API is a set of procedures, routines, tools, or methods that allow a system to connect with another system, either to receive or send data.

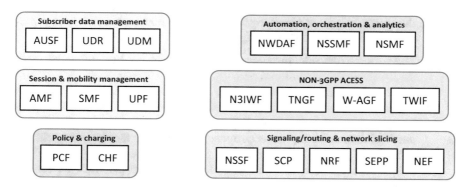

Fig. 9.8 Function aggregation in 5G Core

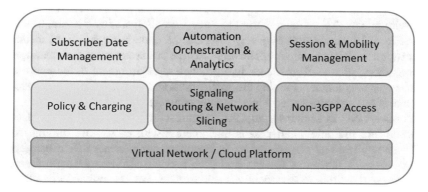

Fig. 9.9 5G multi-vendor CORE

The Core may include ancillary functions to enable increasingly reliable services, notably in the area of routing, security, and for optimizing content transport, as these options emerge in the marketplace.

As stated before, traditionally both hardware and software were provided by a single company and the whole system was proprietary. This makes the system plastered. In this case, increasing the capacity of the network always implies exchanging equipment. With this new technology, it was possible to introduce new network concepts that make operators' life easier and reduce both CAPEX and OPEX costs. These innovations are listed below:

1. *Network Virtualization* allows all Core functions, to be implemented in virtual machines (VM). These machines run on an open-source operating system, usually LINUX type. Between the operating system and the virtual machine, there is a cloud layer, called Hypervisor, through which several virtual machines can have access to shared hardware resources such as: CPU, memory, network, storage, etc. These virtual machines can also run with proprietary operating systems, and usually, in these cases, the applications used are also proprietary.

2. *Cloud Native*, the system is designed to use technologies developed for cloud applications, allowing, for example, the adoption of a fully scalable strategy, so that the system can grow according to demand in a manner absolutely transparent to the user (operator). This strategy implies:

    (a) The apps are agnostic, meaning the operator does not have to worry about the hardware the app will run on.
    (b) In general, applications that run in the cloud are small and controllable.
    (c) Since the processing is distributed, the user (operator) does not have to worry so much about the global operation of the system, because in case of need of maintenance, or occurrence of failures, the impacts will be minimized.
    (d) Automation and orchestration (high-level management) software can be used to ensure the system functions.

3. *Containers*, is a software application composed only of executable code associated with all the dependent requirements. They can run in various computing environments. All containers use the same Kernel/Operating System, as shown in Fig. 9.10. Again, this is a way to share the underlying hardware resources between various container applications. There are systems used to manage the containers, such as Kubernetes. In a production environment, it is important to manage the containers running the applications and ensure that there is no downtime. If one container goes down, another container needs to be started.

4. *Microservices,* in a microservices architecture, complex applications are decomposed into multiple smaller parts that interact with each other through APIs. These applications are created as independent components that execute each process as if it were a service, and each service performs a single function. Because they run independently, each service can be individually updated, deployed, and scaled to meet the demand for specific functions of an application.

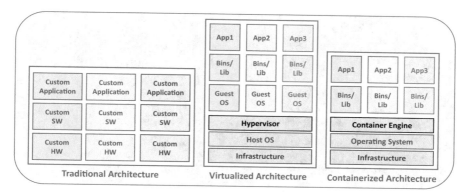

Fig. 9.10 Evolution of the CORE Architecture

5. *Automation and Orchestration*, with the continuous growth of traffic demand, an increase in the number of network elements, the insertion of new services and more complex functions that require specific parameters, the network begins to require more and more control and supervision. It is almost impossible for network management and service orchestration to be done without the aid of automation. In fact, the use of techniques such as *AI (Artificial Intelligence)* and *ML (Machine Learning)* will be incorporated at all levels of the system bringing operational improvements such as greater visibility, better performance, dynamic traffic control, and easy network resizing, besides enabling maintenance in "zero time." The massive use of new technologies also reaches the access network (RAN) through the so-called *SON – Self-Organizing Network* which, in literal translation, is a network that has the ability to self-organize. Using artificial intelligence, this network can automatically plan, configure, manage, optimize, and repair itself without requiring human operators. Organizations such as 3GPP (3rd Generation Partnership Project) and NGMN (Next Generation Mobile Networks) are committed to the development of SON.

Figure 9.11 shows us the complete architecture of the virtual network NFV (Network Function Virtualization).

The main functions of a virtualized network are:

1. *VNFI – Layer*, which is Virtual Network Function Infrastructure Layer, basically composed of hardware resources such as processors, storage, and network, which are virtualized and shared by the upper layers.
2. *VNF Layer*, which is essentially the layer where virtual network functions will be implemented, for example, EPC Core functions (MME, HSS, etc.). In this same layer additional functions can also be installed, such as a virtual DHCP (Dynamic Host Configuration Protocol) server, enabling IP address control, Firewall, Gateway, etc.

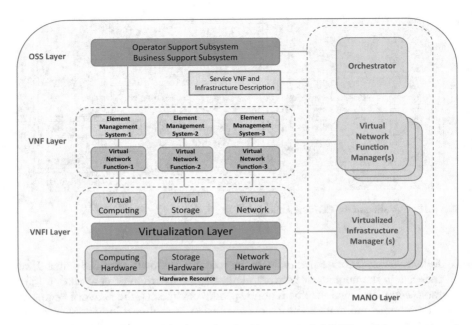

Fig. 9.11 Network Function Virtualization Architecture and MANO – Management and Orchestration

3. *OSS/BSS Layer*, the OSS (Operation Support Subsystem) that takes care of the network management, which mainly comprises the management of failures, configurations, and services. The BSS (Business Support Subsystem) Layer, on the other hand, takes care of business, customers, product management, etc. The OSS/BSS environment provides essential business functions and applications, such as operations support and billing driving the convergence of IT and telecommunications. This convergence will renew network operations and enable new services and business models in current and future networks.

4. *MANO Layer*, which is the Management and Orchestration layer that interacts with the other components managing the infrastructure resources allocates and control VNF resources. In addition, it collects information of the performance of the set, monitoring the capacity and optimization of available resources. It monitors the KPI (Key Performance Indicators) to ensure that the objectives are being achieved.

At some point in our narrative, we mentioned that 5G would open an opportunity for network management to be used more effectively and profitably. After having described concepts such as Network Slicing and Network Function Virtualization, it is easier to demonstrate how this can become a reality, making the network operation much more profitable, creating conditions to better serve customers, and opening new business opportunities.

The concept of *MVNO* (Mobile Virtual Network Operator) is not new and has already been used, but now with 5G, this operating mode can be explored more intensively. A mobile virtual network operator or MVNO is a wireless communication service provider. Unlike traditional mobile network operators MNOs (Mobile Network Operators), MVNOs do not own the infrastructure they use. Instead, they lease resources from an MNO and use this infrastructure to provide a unique set of services to their markets. MVNOs purchase access to a traditional operator's services at discounted rates or wholesale prices and set retail prices for sale to consumers.

CORE's virtualization allows two or more completely independent operators to be housed in the same infrastructure, in a completely secure way, within all the necessary requirements. This way, you can adopt other ways to explore the network. Investments and operating expenses can be treated together, enabling new deployments and the creation of new forms of revenue. Even scarce and essential resources such as frequency spectrum can be used much more efficiently by operators. The greater programming and control capacity offered by network virtualization opens new possibilities for sharing hardware and software, as well as enabling the creation and use of new forms of business and commercialization of services.

Besides the MVNOs, there are also the *MOCN* (Multi-Operators Core Network). The operational model of the MOCN includes a central element, between the RAN and the Core Networks of several operators, called Open RAN aggregator, which has the ability to intercept traffic and route the 2G, 3G, or 4G to the appropriate Core. This allows RAN sharing to happen without complications. The software platform simply requires connections to each Core and handles the more complex job of routing traffic appropriately. In turn, each Core Network manages its users as if they were on the home network. This allows services such as RCS (Rich Communication Service), VoLTE (Voice over LTE), LI (Lawful Interception), etc. to remain viable regardless of whether, or not, the User is active on the operator's local network.

There are also *MORAN* (Multi-Operator RAN), where two, or more, operators share only the access network (RAN). MORAN is used mainly for coverage expansion with the lowest possible investment in hardware.

Another alternative is the *GWCN* (Gateway Core Network), a variant of MORAN in which the RAN is common, and the Core EPC is dedicated to each operator, but the MME (Mobility Management Entity), is shared.

There are specialized companies, which are organized to enable these models, they are the *MVNE* (Mobile Virtual Network Enabler). These companies provide support to a mobile virtual network operator (MVNO), taking care of administration, operation support systems, and defining the entire business infrastructure, seeking additional services to improve the financial profitability of the back-office, participating in the definition of prices to be charged, and the choice of distribution and promotion channels.

The innovations of 5G do not stop, history repeats itself once again, and it is worth remembering, again. Around the 1970s, all data processing was centralized, and users shared the "electronic brain," the mainframes, using so-called "dumb" terminals, devoid of any capacity of their own. Some manufacturers launched mini-computers, the mini mainframes, which had an advantage, besides being more advanced and cheaper, they had the ability to communicate, performing tasks such as message routing when networked. They used several protocols, which evolved until they reached the current standards such as TCP/IP (Transmission Control Protocol), and UDP/IP (User Datagram Protocol) which are used extensively, including, as shown, in 5G.

The advent of the PC (Personal Computer), that at first was thought to be for domestic use, quickly obsoleted the monolithic and proprietary computing architectures, whose manufacturers disappeared in less than 10 years. Along with PCs came the application programs, such as Word, Excel, PowerPoint, and many others, that became indispensable. Most importantly, PCs communicate with each other and thus allow decentralized processing, which is much faster, more confidential, and efficient, at a much lower cost, socializing data processing. Gradually the need to share files began to emerge, which were placed on servers, and with them countless tools that allow sharing to be safe, confidential, and efficient. Also, the personal backups started to be stored in special servers that were placed "in the clouds."

Something similar occurs with 5G. With the increase in traffic, which overloads fronthaul, midhaul, and backhaul, besides the identification of problems, such as latency, that could make most applications unviable if they could not be circumvented or minimized, it was necessary to restructure the entire existing system. For this, several techniques already employed in the computing environment were reused in the telecommunications environment. The telephony network went through an evolutionary phenomenon like what we saw in the past with computers, where processing had to be decentralized, and the systems had to be simplified and adapted to better meet the needs of the consumer market. Thus, was born the *Edge Computing* that focuses on providing more and better services to the customer, improving the quality of user experience reaching a giant market in need of new technologies. The network had to adapt. In this scenario, obviously, it would not make sense that much of the traffic, data packets generated, for example, by IoT applications, had to travel miles to be processed and then return at a time that, most likely, the result of the processing would be no longer relevant. This occurs with various applications such as Drones, robots, VR (Virtual Reality), AR (Augmented Reality), Games, and even more complex applications, such as autonomous cars. It is estimated that in a few years, up to 75% of processing can be done on the edges of networks and that there will be more than 35,000 Edges around the world by 2024. Edge computing offers the extra advantage of relieving traffic in the network center because within this model, only the result of the processing will be reported to the central data centers.

Obviously, the transfer of processing to the edges could only have occurred with the cheapening, the integration and the possibility of having a robust and programmable hardware, able to run all the necessary applications. That is, all this only became possible thanks to the development of technologies such as *GPU* (Graphics Processing Units) (graphics accelerators), *ARM* processors Advance RISC (Reduced Instruction Set Computer) Machines) (which allows the processing of large volumes of data), and even the already known *ASICS* (Application Specific Integrated Circuits) intended for the processing of specific functions.

The main advantages of EDGE Computing are summarized as follows:

- Reduces latency.
- Minimizes traffic on the transport network.
- It increases the security of information, which is less exposed and circumscribed in a smaller geographical area.
- Improves user perception of the quality of service provided, i.e., improves the so-called *QoE* (Quality of Experience).
- Helps the operator to reduce the volume of investments and operating costs. The *TCO* (Total Cost of Ownership) is expected to be reduced by creating a cheaper and decentralized infrastructure.
- Increases network reliability, by distributing content among the various Edges and mini–Data Centers.
- It creates opportunities for the emergence of companies focused exclusively on providing specific services.
- Smaller computing resources that take up little space and cost less are now being distributed and installed in a wide variety of places such as shops, cafes, bars, restaurants, etc.

This entire evolution, or rather technological revolution, initiated with 5G brings with it new business models that range from the sale of automation services to the opening of new markets for fiber optic and infrastructure installers, mobile phone operators, IT system providers, cloud providers, and a whole wide chain of related services. These services can be offered by independent and specialized companies.

Chapter 10
The Evolution of RAN (Radio Access Network), D-RAN, C-RAN, V-RAN, and O-RAN

We are coming to one of the final chapters of our book. We saw how the evolution of cellular systems happened with the progressive transformation of the various architectures and how the forms of digital modulation allowed to add more capacity to the systems. We had the opportunity to analyze in detail the massive MIMO antennas, which are arrays of high gain antennas capable of meeting, almost miraculously, the high-quality standards required by 5G, in addition to working with massive densities of connections and providing very high throughput per user.

Our knowledge has deepened to the point of describing the phenomenon of the "softwarerization" of cellular technology, in which virtualization and Cloud techniques, previously used only in computational applications, have been applied to telecommunications networks adding agility and flexibility to processes. Proprietary hardware is giving way to computing platforms of generic use, which employ *COTS* (Commercial Off-The-Shelf) units, easy to install and replace and which, loaded with the appropriate software, allow the virtualization of important RAN functions and applications using native Cloud principles. The other hardware, that is used for processing the radio frequency signals, is also becoming almost an off-the-shelf item thanks to the huge advances in semiconductor technology. This is exactly what motivated us to name this book *The Architectural and Technological Revolution of 5G*.

5G is ultimately the result of the convergence of several technologies and the ongoing development of both hardware and software. It is a reality that is coming to fruition, thanks to an open and participatory working approach that has enabled stakeholders, i.e., users (operators) and manufacturers, to define network needs jointly within the 3GPP framework. The standardization of interfaces and the detailed description of network functions have opened the market for new vendors to develop their own solutions and generated opportunities for many technology start-ups.

Although we have already talked about the RAN (Radio Access Network) previously, elements were missing to show how the protocols necessary for the execution of the functions fundamental to its operation can be dismembered in order to optimize, both economically and operationally, its deployment.

© The Author(s), under exclusive license to Springer Nature Switzerland AG 2023
J. L. Frauendorf, É. Almeida de Souza, *The Architectural and Technological Revolution of 5G*, https://doi.org/10.1007/978-3-031-10650-7_10

Let's first evaluate the motivating factors for this whole revolution, which are many and diverse. There is no single answer, but rather solutions that must be implemented according to the dynamic needs of the market.

Probably the main factor is the growing demand for greater individual capacity per user. This implies cell densification and a sharp increase in traffic volume in the various hauls, the data transport routes between the data modules and the Core that controls the network. Just to give you an idea, the traffic volume of a cell operating with a 64-element Massive MIMO antenna with 100 MHz bandwidth can reach more than 300 Gbps or exceed 3000 Gbps in special cases. This volume of data traffic will continue to grow dramatically as more massive MIMO antennas are used and the availability of large spectrum bands is becoming available to operators, particularly in systems expected to operate in the millimeter-wave bands. Table 10.1 details these claims, based on theoretical values.

The second factor is latency, as discussed earlier. Some new applications can't be deployed as long as the traditional 10 ms in 4G, the time lapse between the round trip of a message from source to destination, can't be reduced. New applications require less than 1 ms.

The third main factor is the number of network "connections." With the likely growth of IoT users and connections served by cellular systems, the trend is for traffic to increase exponentially in the coming years. Estimates show a staggering number: more than 1000 IoTs/km^2. In addition, there is the huge challenge that an IoT device can run for up to 10 years without a battery change. This causes its RF signal emissions to work at an extremely low level, transmitting or receiving signals for very short periods.

There is also a fourth challenge: communications must be 99.9999% reliable in some cases!

Recalling how the RAN evolved:

Table 10.1 Transport network requirements - 5G Fronthaul

Number of Antennas	Available Bandwidth			
	10MHz	20MHz	200MHz	1GHz
2	1 Gbps	2 Gbps	20 Gbps	100 Gbps
8	4 Gbps	8 Gbps	80 Gbps	400 Gbps
64	32 Gbps	64 Gbps	640 Gbps	3,200 Gbps
256	128 Gbps	256 Gbps	2,560 Gbps	12,800 Gbps

Data extracted from Annex A 3GPP TR 38.801

D-RAN (Distributed RAN), whose distributed architecture requires the RRU (Remote Radio Unit) and BBU (Base Band Unit) to be collocated. The cell connects to the Core through the backhaul, as shown in Fig. 10.1.

C-RAN (Centralized RAN/Cloud RAN), where RRUs are distributed and BBUs are centralized. RRUs and BBUs connect through the fronthaul, and BBUs and the Core connect through the backhaul, as shown in Fig. 10.2. C-RAN is often also called Cloud-RAN due to the application of Cloud technologies in the telecommunications environment.

V-RAN (Virtual RAN), where most functions are processed virtually, even in proprietary solutions.

O-RAN (Open RAN), a solution that emerged as part of 5G.

Before we get into the subject of O-RAN, let's talk about an innovation that occurred with the separation of BBU functions into two distinct and geographically separated units: the *DU* (Distributed Unit) and the *CU* (Centralized Unit). Now that we are familiar with the architecture of the RAN in cellular systems, we will see how this separation was made so that the processing normally performed in distinct protocol layers occurs in physically distinct units. Let's recall the 5G protocol stack, Fig. 10.3, while identifying which layers have been allocated in each unit. It is important to note that the PHY layer has been broken into two parts – the *High PHY* and the *Low PHY*. This will occur in other layers as well, as we will have the opportunity to observe later.

The *RUs* correspond to the *Radio Units* and all the digital processing of part of the physical layer, the Low PHY, which includes the processing of functions such as massive MIMO and Beamforming. The biggest concerns with these units are about size, weight, and power consumption.

Fig. 10.1 D-RAN (Distributed RAN)

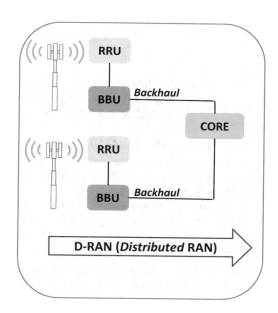

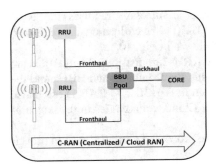

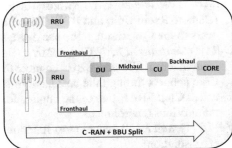

Fig. 10.2   C-RAN (Centralized RAN)

Fig. 10.3   5G Protocol
Stacks

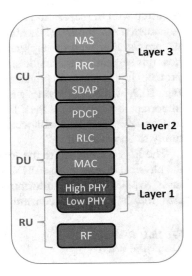

The *DUs* are the *Distributed Units*, which should not be far from the RUs. The normal distance is between 1 km and 20 km, interconnected by fiber optic. Several RUs may be connected to the DUs. The DUs connect to the RUs through the fronthaul.

The *CUs, the Centralized Units*, take care of the rest of the RAN functions. The CUs can be connected to multiple DUs, which in turn can support multiple RUs. The DUs connect to the CUs through the midhaul, while the CUs connect to the Core through the backhaul.

The reason for the separation of BBU in DU+CU is due to an important factor. While the DUs concentrate most of the critical functions with respect to timing, which must be executed in real time (Real-Time Functions) and are part of Layer 1 and part of Layer 2, where the scheduling function is allocated (management of the sequencing of data packets transmission), the CUs concentrate the functions that are not critical with respect to timing, functions that are processed in part of Layer 2 and

Layer 3. Therefore, the DUs, by being closer to the RUs, can better control latency. The management done by CUs allows a better distribution of traffic between the RUs.

On the other hand, the separation of DUs and RUs is due to three distinct facts: (1) reduction in the cost of the RUs, since less intelligent RUs cost less; (2) possibility of managing several RUs from the DU, which facilitates the implementation of the *CoMP-Coordinated Multi-Point* model, in which, under the assumption of the existence of several cells that have MIMO/massive MIMO and Beamforming antennas, it allows coordinating the action of several antennas, avoiding interference from one system in another and making better use of spatial diversity to improve the coverage of a region (the DU can choose which antenna from the set of antennas can better serve a user); (3) the resources allocated to a DU can be shared by several RUs.

With Cloud technology available, it is not hard to imagine that many of the solutions can count on high levels of function virtualization and associated cost gains in both investment and operating expenses.

The natural evolution was the creation of the *Open RAN Alliance*, which brings together equipment manufacturers, SW, and solution developers along with operators, who came together to, based on the technological advances already achieved, be able to diversify the ecosystem of solution supply for cellular systems and, in particular, 5G. The differences between the various RAN generations are summarized in Table 10.2.

The model adopted by O-RAN allows total independence to compose the network with solutions from different software and hardware suppliers, which, thanks to standardized interfaces and protocols, can be interconnected to the system, allowing the operator to choose the appropriate suppliers for the access network it intends to install, as shown in Fig. 10.4.

With the advent of O-RAN, possible alternatives for disaggregation of the functions performed by the RAN went beyond those previously defined (RU/DU/CU) and provide up to eight distinct options, shown in Fig. 10.5, although only part of them have been considered feasible for possible implementation. They are Options 2, 6, 7.2x, and 8.

Table 10.2 Different Radio Access Network generations

	Baseband Hardware	Baseband Software	Radio Hardware (RRU)	BBU/RRU Interface (Fronthaul)	Interability
C-RAN (Centralized)	Proprietary Technology	Proprietary Software	Propietary Hardware	Proprietary Interface	Radio + BBU (HW + SW single suplier)
V-RAN (virtualized)	COTS	Proprietary Software	Propietary Hardware	Proprietary Interface	Radio + BBU (SW single suplier)
O-RAN (disaggregate)	COTS	Software Open Interface	COTS	Open Inteface	Radio + BBU (HW + SW various suplier)

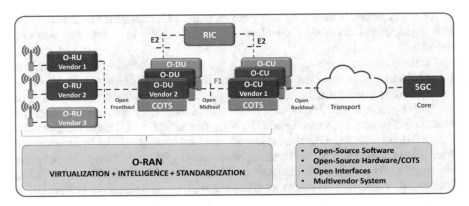

Fig. 10.4 Different Open RAN providers

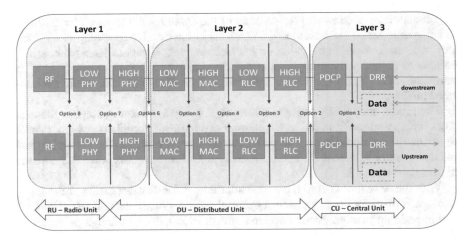

Fig. 10.5 5G Functional Split Options

Based on this plurality of models, both the O-RAN Alliance and the Small Cell Forum, another institution that seeks to enable smaller and low-cost cells aimed at complementing larger solutions, proposed options for fractioning BBU services. The two institutions chose two close alternatives, Option 6 and Option 7.2x. In Option 6, there is no unbundling of Layer 1 PHY, while in Option 7.2x, Layer 1 is split into Low PHY and High PHY, as shown in Fig. 10.6.

Figure 10.7 shows the most viable options, Options 7.2x, 6, and 2, along with the more traditional Options 1 and 8. Again, it is important to remember that only Layer 1 and Layer 2 can be RU/DU, and Layer 3 is always part of the Central Unit (CU).

Let us analyze in detail each of the options, remembering that the higher the separation of functions (higher layers) in the protocol stack, the lower the split ranking. This particularity often confuses the reader when it is not clearly explained what is considered as high or low, i.e., whether the observation refers to the

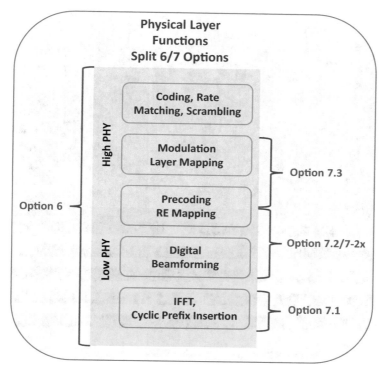

Fig. 10.6 Low PHY and High PHY Split for Split Options 6 and 7

classification of the split 1....8 or to the aggregation level of the functions. Figure 10.8 helps us understand this fact.

Let's clarify the Options, one by one:

OPTION 1 (Split 1) is the typical version of a traditional 4G RAN or small cell in which both RU, DU and CU are practically integrated. The advantage is that the traffic volume with the upper layers of the L3 layer is small. In some cases, this solution can be interesting in edge computing situations, or when the latency requirement is low, or when the user data needs to be located close to the transmission unit. It tends to be a complex, expensive, heavy solution, which occupies a lot of space and consumes a lot of energy. This option is hardly mentioned in the literature in general.

OPTION 2 (Split 2) is the option that concentrates the largest number of functions, and, among all the models, it is considered to be the most viable. It is the reverse situation of Option 8, as we will see later. In Split 2, almost all functions are more concentrated. For this reason, fronthaul requirements can be a little more "relaxed," both in traffic volume and latency. On the other hand, it is the solution that tends to be more expensive, takes up more space, and consumes more power. This model makes the use of *CoMP* (Coordinated Multi-point) technology practically impossible, since it is not able to share functions that allow the use of this

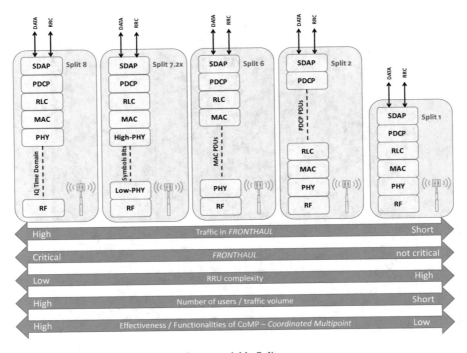

Fig. 10.7 Main differences between the most viable Splits

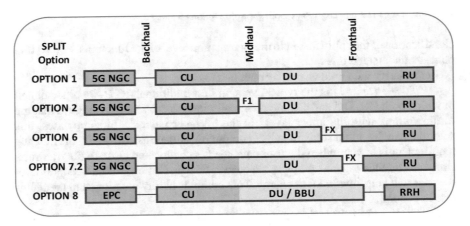

Fig. 10.8 Split options and Fronthaul / Midhaul / Backhaul connections

technique. It can be an interesting solution for fixed networks *FWA* (Fixed Wireless Access).

Because Option 8 is opposed to Option 2, we will address it before the others.

Option 8 (Split 8) corresponds fundamentally to the C-RAN model, but with an open architecture, in which all the higher functions of the 'stack' are centralized, thereby obtaining all the benefits of sharing functions and allowing better distribution of traffic between cells due to the functions of the various RUs being shared centrally. It also allows the virtualization of a large number of functions, which are now processed in COTS servers. The RUs should cost less because they are less complex, lighter, take up less space, and consume less energy. Updates and upgrades require less effort from operators and demand fewer site visits. On the other hand, they demand a lot from the fronthaul, both in terms of traffic volume and latency. This is the ideal situation for the use of CoMP. Another relevant factor: it can be used as a solution to aggregate all previous generations (2G, 3G, and 4G), since each one is occupying its own operating frequency band. The distance between a RU and the DU/CU should be limited to 20 km, considering a fronthaul that meets the requirements of the applications demanded by the market. There is a need, obviously, for a fiber optic type connection for this interconnection. To meet this requirement, the former CPRI evolved and the eCPRI (enhanced Common Public Radio Interface) appeared in its place. This allows the aggregation of traffic from several RUs and facilitates migration from 4G to 5G.

Table 10.3 gives an idea of the range of distances that can be used with each haul:

It remains for us to discuss the other two options considered by operators, options 6 and 7.2.

Option 6 (Split 6) is an option in which all physical layer processing (PHY) is restricted to the RU, which means a drastic reduction in the traffic requirement at the fronthaul. Table 10.4 demonstrates this fact. The drawback of this option is that it limits the opportunity to apply a technique called pooling, which also allows for the joint utilization of computational resources by several RUs. In this configuration, pooling is restricted to the Data Link Layer and the Network Layer, which can mean a gain of only 20%, and the remaining 80% refers to sharing the Physical Layer processing. This option was chosen by SCF (Small Cell Forum) as the ideal for smaller cells.

Option 7.2 (Split 7.2) is the most popular option among O-RAM Alliance operators as it is a good compromise solution that enables large cell capacity, high reliability, use of relatively simple RUs with size, weight, and power consumption that support network densification and at the same time sharing by many operators of a single infrastructure as it is characterized as a neutral network. DU and CU can be together, and their functions are processed by a single server, which in turn can

Table 10.3 Typical Distances from Fronthaul/Midhaul/Backhaul

	Typical Distance
Fronthaul	1 - 20 km
Midhaul	20 - 40 km
Backhaul	20 - 300 km (CORE)

Source: **ITU**

Table 10.4 Transmission rates required by the DL/UL and respective latencies according to the Split option

Split Option	Downlink Rate	Uplink Rate	Latency (each direction)
Option 1	4 Gbps	3 Gbps	1 - 10 ms
Option 2	4016 Mbps	3024 Mbps	1 - 10 ms
Option 6	4133 Mbps	5640 Mbps	100 – 500 µs
Option 7.2	10.1 - 22.2 Gbps	16.6 - 21.6 Gbps	100 – 500 µs
Option 8	157.3 Gbps	157.3 Gbps	< 100 µs

Data extracted from Annex A 3GPP TR 38.801
Calculated for an RF configuration = 100 MHz / 256-QAM / 8 MIMO Layers / 32 Antennas

serve multiple RUs. This option can take full advantage of massive MIMO, which relies heavily on the interaction between multiple gNodeBs. This version uses the aforementioned eCPRI, compatible with an Ethernet-like protocol, which makes it ideal for use in urban areas and indoor environments such as factories and office blocks, places that require 5G technology. The use of cloud solutions is also fundamental for the economic aspect. The disaggregation of RANs brings greater flexibility and allows the arising of companies specialized in breaking with the tradition of fully verticalized operations. This option is a good compromise between centralization and decentralization of processing functions, while allowing the use of innovations such as Carrier Aggregation, MIMO, and CoMP. It is also worth saying that Option 7 is presented with three different versions, which differ only by the volume of data in the fronthaul, and version 7.2 is the one that presents greater acceptance among operators.

At this point, it is important to keep in mind that most likely a system based on a single alternative, which includes mainly the frequency of operation of the system, will not be sufficient to meet market demands. The model of SK Telecom from South Korea is a good example. They use a lower frequency, 3.5 GHZ, to solve the coverage problem, and 28 GHz to meet a large demand for localized traffic that requires low latency. Figure 10.9 is based on data released by SK Telecom itself.

The division of BBU functions into RU, DU and CU allows services with different latency and processing capacity requirements to be served but does not fully address latency reduction needs. In the traditional system, services are connected at the edge of the network and processed in large data centers, which are located in the network center or Core. Typically, the distance travelled by the data/voice packet from the network edge to the Core can reach hundreds of kilometers. Consequently, the travel time of the packet is also very large.

One way to minimize the problem was to reduce the distance between the service and the processing center. This gave birth to the concept of *Edge Computing*, which

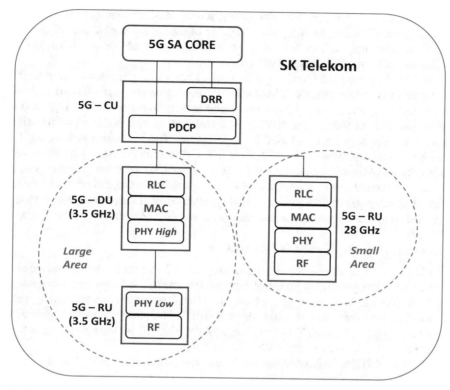

Fig. 10.9 SK Telecom model for 5G coverage area

involves building mini data centers near the RUs installed at the edges of the net-work, an area known as Edge or Fog. This allows applications to be processed as close as possible to the point where the service will be consumed. With this, net-works gain near real-time processing capacity, a key factor for the development of new services.

The new applications require integration between the knowledge areas of IT (information technology) and telecommunications which, in this case, lead to the creation of the *MEC platform or Multi-access Edge Computing/Mobile Edge Computing)* in order to design the computational and radio communication struc-ture required to offer services at the network edge, as we will see below.

ISG (Industry Specification Group) is a working group within *ETSI* (European Telecommunications Standards Institute) that aims to create an open and standard-ized environment that enables the efficient and seamless integration of a range of entities from the world of telecommunications, IT resource providers, and Cloud computing within the RAN (Radio Access Network). In 2012, the ETSI ISG group created the term MEC (Mobile Edge Computing) in order to emphasize the need for standardization in Edge Computing. The ISG MEC specifies the elements required to enable applications to be hosted in an edge computing environment in which solutions are the result of the joint interaction of multiple vendors and multiple access types. ETSI has itself renamed MEC to Multi-access Edge Computing in an effort to accurately reflect the focus on multiple types of access technologies, including LTE, 5G, Wi-Fi, and fixed access.

MEC (Multi-access Edge Computing) brings computing resources closer to the end user and can be defined as Cloud services running on the edge of a network and dedicated to specific tasks. Aimed especially at very low latency and high band-width services, as well as real-time or near-real-time access, it allows information from the radio network to be used by applications. The MEC is transforming the topology of mobile networks, which were initially designed to work with voice and data, into an application platform for the new services. As we know, 5G is expected to support several use cases, including Cloud operation, remote control of robots, Virtual Reality games, Augmented Reality, predictive maintenance, autonomous vehicles, and numerous IoT automation and sensing applications, among other services.

Briefly, the main objectives of the MEC are:

- *Gain in Response Time* by reducing latency, a key factor for functions that require real-time analysis, such as facial recognition, robots, IoTs, interactive games, *V2X* (Vehicle-to-Everything), which includes all vehicle communication with external elements such as traffic signs, traffic lights, other vehicles, Augmented Reality, Virtual Reality, Telemedicine, in short, virtually all the new services that are expected to be available with 5G.
- *Decreased Traffic Volume in the Network Center*, since much of the traffic will be handled at the edges of the network, there is resource release in the network center, so traffic is better organized and processors in the network center work with a lower load.
- *Increased Security*, which aims to reduce the occurrence of failures caused by attacks, and reasons for interruptions in data transport networks, since much of the processing is performed in a distributed manner, thus generating greater autonomy and increasing security against possible invasions by hackers.
- *DDoS* (Distributed Denial of Service Attack) is a process in which malicious people (hackers) generate extraordinarily heavy traffic on the network so that it can no longer operate normally. Again, distributed processing tends to minimize the occurrence of these types of attacks.
- *Interoperability*, which allows Edge Computing to work as an interface between legacy or specialized systems, enabling the use of special APIs to communicate with more modern machines and/or devices.
- *Infrastructure scalability*, which can be gradually expanded as local demand increases.
- *Increased Offering and Quality of Service (QoS)* to users.

As already seen in other situations, decentralization always aims, besides the factors listed above, reducing CAPEX and OPEX.

It is important to remember that Edge Computing is not a standalone system. It must be connected to the main data center, with which it is in constant interaction and where important information is stored and must be maintained centralized, although Edge Computing can be seen as a set of micro data centers or Cloud Edges, which store, and process data generated locally by IoT devices or even robots. Video and program content most accessed by users in each location can also be stored locally in micro data centers.

We cannot end this chapter without looking back at some of the innovative technologies that are emerging within the new generations of access networks, LTE/5G, which are all directly or indirectly related to the items discussed above.

CLOUD, general concept of cloud computing that is nothing more than on-demand access to computing resources, applications, servers (physical and virtual servers), data storage, development tools, and network resources that are hosted in a Data Center (DC). This DC can occupy a single site or have several sites operating in a shared and redundant manner. The "cloudification" or application of cloud services in the telecommunications network allows the use of resources such as virtualization in the communications network, making it more agile, flexible, and scalable. Cloud technology also makes it possible to control and enable traffic in a network where there are millions of connected nodes, providing higher speeds, lower latency, and greater capacity. Thus, operators that are preparing their networks for 5G will be able to quickly deliver innovative new services, meet growing customer expectations and maximize revenue. We can mention some key technologies that will be used in the 5G Cloud:

Virtual Networks (VN) allow you to run on virtual machine (VM) functions previously performed by dedicated devices within the network architecture. Using COTS servers and software that enables network functions virtualization (NFV), the operator can install/remove new elements easily and quickly, at the lowest cost – in addition to exploiting the features normally used by VMs, such as scalability (up/down), adjusting the network size according to real-time demand, performing online software updates without interruptions or even moving the VM to another server in case of maintenance. In other words, it streamlines system operations, all through software.

Network Slice Instance (NSI) allows "virtual network slices" to be assembled according to the specific needs of a service (such as latency, throughput, etc.), meeting industry or individual service requirements, and virtually separating these service flows within the network in order to maintain the desired quality of service (QoS) for each type of service and allow for its direction and control within the network.

Network Slice Management Function (NSMF) are 5G self-managed networks that enable orchestration and dynamic adjustment of the network in order to maintain performance at an optimal level. This is made possible by employing artificial intelligence/machine learning (AI/ML), which enables rapid provisioning decisions generating changes in the operating conditions of the 5G network and dynamically maintaining the performance of each slice within specific quality levels.

Software Defined Radio (SDR) are new radios, which can be fully software configured such that a single platform can be employed in numerous applications and in which the configuration of radio functions can be performed remotely. It can be reconfigured and updated at any time as required by the operator. Thus, a single platform can be configured to operate as UMTS, HSPA, LTE, or 5G without hardware changes, even though the modulation scheme or operating frequency of each of these generations is completely different.

Software Defined Network (SDN) enables operators to quickly manage and reconfigure the resources used in the network through automated provisioning and policy-based management. These help operators respond to fluctuations in resource and traffic demand and ensure their efficiency. A software-defined network consists of three main components, which may or may not be located in the same physical area:

- *Applications*, responsible for relaying information on the network, requests for availability or allocation of specific resources.
- *Controllers*, which act as load balancers and communicate with applications to determine the destination of data packets.
- *Network devices*, such as routers, that receive instructions from controllers on how to route packets.

Self-Organizing Network (SON) is an automation technology designed to enable self-configuration, automatic fault management, and self-healing of mobile phone networks. Its functionalities have been specified by organizations such as 3GPP (3rd Generation Partnership Project) and NGMN (Next Generation Mobile Networks). LTE was the first system to benefit from SON functions that allow, for example, newly added eNodeBs to self-configure themselves online, while all other operating eNodeBs connected to it will be able to automatically update their configurations and parameters in response to the insertion of the new Base Station. They will also be able to change their configuration automatically based on algorithms capable of responding to changes in network performance and/or observed radio conditions. In addition, self-correcting mechanisms may be triggered to temporarily compensate for a detected equipment failure while awaiting a more permanent solution. Although the SON function was created to enable access network optimization (RAN), operators are expanding its scope with a focus primarily on the integrity of the network, incorporating dynamic optimization and automatic troubleshooting functions, applicable even while the network is in operation. In addition, artificial intelligence and deep learning techniques are increasingly being used in network control and maintenance. In addition, automation functions have been included in other parts of the network, including the Core itself and the data transport networks, fundamental functions due to the explosive increase in the number of elements in the network which, as in the case of 5G, makes it difficult or practically impossible to maintain and configure the network in manual mode.

Dynamic Spectrum Sharing (DSS) is a technology that allows two systems to share a spectrum band at the same time. For example, the 4G band can be shared between 4G and 5G in an integrated way. This means that if a user is in an area with 5G DSS coverage and equipped with a device that only has 4G, the system will provide services compatible with the technology it has, 4G. But if he is equipped with a 5G terminal, he will also be able to connect by sharing part of the available spectrum band (bandwidth), because DSS allows to dynamically allocate fractions of spectrum bandwidth to 4G or 5G, depending on the demand of the cell. DSS technology was already known before the introduction of 5G and used to solve local spectrum usage problems, allowing sharing, for example, the 2G spectrum band with 3G or 4G. However, it has become an important technology for use in cases of refarming (relocation) from 4G to 5G or rapid installation of the 5G network, in which DSS stands out as a fast and cost-effective way of installation. Note, however, that 5G DSS does not have the same performance as pure 5G, so the use of 5G DSS should be preceded by a good network planning and verification of the desired performance levels.

RAN Sharing is another concept created to optimize and increase the efficient use of network resources. The objective is to allow the sharing of access networks by several operators. Sharing network infrastructure and adopting virtualization techniques can help an operator reduce significant CAPEX and OPEX values. There are two possible ways of sharing mobile network infrastructure:

- Passive sharing, limited to passive network elements such as radio antennas, power supplies, cabinets, towers, security alarms, etc.
- Active sharing, which includes transport infrastructure (fiber, cables, etc.), baseband processing resources, and potentially radio spectrum.

Due to the advances in 5G, there is now talk of sharing even the Core Network.

Multi-RAT (Multi-Radio Access Technologies) combines various access network technologies to enable the most diverse services, whether GSM, UMTS, LTE, 5G NR, or Wi-Fi, in a transparent manner for the user. In this concept, many suppliers develop the so-called *Multi-band RRH* (Multi-band Remote Radio Head), which supports multiple frequencies and multiple technologies, already mentioned above (GSM, UMTS, LTE, or 5G NR), causing operating costs to decrease by reducing the complexity of the sites, by saving space and by saving energy demand.

Multi-band, Multi-RAT, and Multi-Operator Antenna System   5G was developed to work basically with Small Cells. However, *Macro Cells*, which have been used to provide cellular coverage for several years, have some important advantages, especially in terms of serving subscribers in geographically dispersed areas and supporting legacy technologies (2G, 3G, and 4G). The fact that this infrastructure is already installed and licensed means that it still has a very relevant role in telecom systems and will probably continue to be an important element in the access network for several years. In this scenario, we highlight the evolution of antennas capable of working with various frequency bands – for example, 700/800 MHz, 1800 MHz, 2100 MHz, and 2600 MHz to support legacy technologies, in addition to adding, more recently, the frequencies of 5G, such as 3.5 GHz. Obviously, these systems are designed to support Multi-RAT access technologies, i.e., they work with GSM, UMTS, LTE, 5G NR, and Wi-Fi technologies. Some models even provide MIMO solutions with multiple transmit and receive antennas (8T8R up to 64T64R).

Because 5G is an architecture based on services and the improvement of QoE, i.e., the quality of the user experience, it will certainly require the installation of a large mesh of small cells. But it is also expected that 4G and 5G technologies will coexist for many years to come. This implies that Small Cells and Macro Cells should operate in the same network, each serving a certain niche of services. At the same time, interference reduction techniques, error correction, and extensive use of Artificial Intelligence and Deep Learning will be required to ensure that these technologies can exist side by side.

Another important issue is the sharing of infrastructure by multiple operators in order to reduce costs, which also tends to increase the level of demand on sites. As 2G and 3G disappear, and then 4G/LTE, 5G will replace these technologies. It should be noted that 5G devices will become more popular and affordable, with the volume of units sold increasing as new applications begin to be used. It is only a matter of time before it becomes the dominant technology.

We have made a point of mentioning all these innovations with a single objective: to show that there are no longer any technological barriers to the "softwarerization" of cellular networks and that technological innovations are bringing effective and highly efficient solutions. They will enable us to upgrade existing systems and adapt the network to meet the new technical quality criteria required by 5G, which in turn supports the innovations of Industry 4.0 and new services designed for a new era.

Chapter 11
Final Remarks

This book has a very special goal: to invite the reader to learn about the technology intrinsic to everyday things, such as the simple act of sending a message by cell phone, watching a video, or holding a meeting online. Questions like "*How does the structure that allows information transport work?*" or "*What is the frequency spectrum?*" or "*How do you trigger a device at a distance?*" allow humanity to reach very far and build a universe that did not even exist in the thoughts of our ancestors. We hope to be able to encourage questioning, curiosity, and the search for knowledge in the mind and soul of the reader. May this be the seed for much more progress and evolution.

These pages were not written by an expert on the subject, but by someone who has worked in the telecommunications industry for decades. As stated at the beginning of the book, the author was impressed with his first readings on 5G and, as a specialized journalist, he began a great deal of research on the subject. Gradually, he put down on paper everything he learned. To enrich the initial manuscript, a second discussion on the topic was held in collaboration with a professional colleague who is also interested in the subject. We also point out that the techniques and concepts used in 5G represent the most advanced technology available today and have very bold objectives, taking network planning, deployment, operation, monitoring, and control to extreme levels, going beyond the limits imposed by existing network architecture and increasing processing capacity and resource exploration in a way never practised before. However, the 5G system is still in its early stages. Progress, discoveries, and innovations have been constant. We hope, with this paper, to draw the reader's attention and pave the way for them to venture out on their own in the search for new knowledge. This is the closing chapter of our work.

Some extra chapters were added in order to enrich the text and recall details and concepts that may be necessary to allow the understanding and acquisition of information that complement and clarify items addressed briefly in the specific chapters.

© The Author(s), under exclusive license to Springer Nature Switzerland AG 2023
J. L. Frauendorf, É. Almeida de Souza, *The Architectural and Technological Revolution of 5G*, https://doi.org/10.1007/978-3-031-10650-7_11

We could not fail to make a quick evaluation and, with that, to synthetically high-light all that was addressed in the initial chapters, at the same time expressing the deepest respect for human intelligence and all that humanity has conquered.

Officially, the *Transistor* was invented in 1947 by Bell Labs Semiconductor Group. This means that all the technological evolution discussed here occurred in the last 75 years. Note that this is a very short period. In other words, technologi-cally, we are still in the Stone Age, building the bases for a future of enormous pos-sibilities. The evolution of the systems happened mainly by means of the mass integration of electronic micro-devices, like transistors, resistors, and capacitors, inside a semiconductor substrate, normally made of silicon, that received the name, very appropriate by the way, of *Integrated Circuits*.

The integrated circuits, in turn, enabled the birth of the *Microprocessor* approxi-mately 50 years ago. More sophisticated, microprocessors gained their own intelli-gence, that is, they worked with a series of digital instructions that allowed them to perform certain functions. A microprocessor can, for example, control all the func-tions of the CPU (central processing unit) of a computer or other digital device. The microprocessor works like an artificial brain. The system can control everything from small devices, such as calculators and mobile phones, to large machines. Along those same lines, the IoT will largely rely on tiny devices capable of perform-ing specific functions, the so-called Systems-on-a-Chip (SoCs), simple, program-mable devices that, used together, can produce quite complex solutions. They are also in vehicles that have in their structure hundreds of SoCs, one with the function of opening the window, another to control the fuel, another to control speed, and so on, automating most of their functions while bringing comfort and safety to the user.

Parallel to all this evolution of electronic circuits, we had the emergence of what was then called the Cellular Phone, which today already transformed into this extremely flexible and powerful structure called 5G. Communication technologies, including wireless, optical fiber, and cables, provide the necessary infrastructure to support other innovative technologies, such as applications used in Industry 4.0 and IoT. It is the combined use of these intelligently combined hardware and software advances that creates the synergy needed to pave the way for an evolution of previ-ously unimagined proportions.

Let's consider relevant points that have enabled the development of mobile telephony:

- *Improvement of modulation technique.* The increase in the volume of transmitted data, or throughput, has become fundamental in serving functions whose operat-ing criteria are stricter, such as 4 K or 8 K video, robot control, among others. One way to increase the amount of transmitted data is by using modulation tech-nique. Digital Modulation is one of the main item responsible for the evolution occurred in recent years and is subject of one of a specific chapter of this book.
- The theoretical limits for transmission have been known since the early 1950s, as a result of the innovative work of *Claude Shannon*. Published in 1948, his article "A Mathematical Theory of Communication" serves as the basis for everything

that exists today in the segment. It shows us how efficient digital communication can be. Note that the North American Claude Shannon was a contemporary of another genius, the Englishman *Alan Turing*, considered the father of Information Theory.

- In the graph shown in Fig. 11.1, we see how high-order modulation rates get closer and closer to the theoretical limit of Shannon Capacity, seeking to obtain an ever-increasing improvement in the transmission rate. High-order modulation schemes increase the capacity of the network. While 4G has a downlink spectral efficiency between 0.074 and 6.1 bits/s/Hz (bits per second per hertz), 5G networks promise efficiencies between 0.12 and 30 bits/s/Hz.
- The figure considers only the capacity up to the 1024-QAM modulation level, but there are equipment in the market working with 2048-QAM or 4096-QAM modulation levels, and the forecast is to reach even higher levels.
- *Antenna development.* Another important factor, especially regarding the improvement of transmission quality, is the evolution of antenna construction techniques. We highlight the use of MIMO, massive MIMO, and Beamforming antennas. These technologies complement the effect of the evolution of the modulation levels of digital signals by using state-of-the-art technology to improve the quality of the transmitted and/or received signal. Received signals can be combined and treated in such a way that they can be recovered even when they

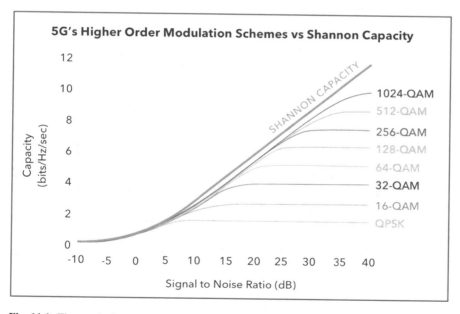

Fig. 11.1 The graph shows the system capacity (bits/Hertz/sec), employing the various modulation levels (from QPSK to 1024-QAM), according to the propagation conditions expressed by the signal to noise level ratio and the threshold capacity established by Shannon. (Source: WAVEFORM)

arrive at receivers at very low levels, considered at first impossible to use because they are immersed in all kinds of noise. Thanks to mathematical processing, it is possible to recover the information. The use of tools such as Beamforming technology enables the quality of the signal provided to the user. The great advantage of a massive MIMO network over a normal network is that it can multiply the capacity of a wireless network connection without requiring more spectrum.

- *Hardware improvement.* The technological advances provided by processors, among which we include the group of dedicated processors, ASICs (Application-Specific Integrated Circuits), memory chips, accelerators, video processors, and GPUs (Graphics Processing Units), which leveraged the use of smartphones, embedded systems, personal computers, and games, enabled the development of "off-the-shelf" hardware, "programmable chips," sets of complete hardware resources that can be programmed by software. The processors also made possible the creation of units for the processing of radio frequency signals, modulators, demodulators, filters, etc. With the introduction of virtualization techniques, the COTS (Commercial Off-The-Shelf) servers for generic use become an important resource to speed up the implementation of systems and reduce installation costs. Software specially developed to emulate network functions can "run" on this "open" hardware, allowing new elements to be installed quickly and easily, which speeds up the execution of network modifications.

- *Massive use of software.* Speaking of software, as far as the cellular network is concerned, we are witnessing an enormous revolution, both architectural and technological, in which various functions are being replaced by software programs and applications. For many years, a restricted group of companies, manufacturers of these equipment, developed for specific use, maintained the hegemony of the market. This is now something that soon will remain in the past. From what has already been said, even legacy structures from previous generations will be replaced by software solutions developed by a universe of vendors that has proliferated in recent years. These solutions will run on top of generic programmable hardware. In addition, the use of virtualization techniques and cloud concepts adds great flexibility to the network, allowing it to adapt quickly to variations in traffic demand. These technological innovations aim to reduce the cost invested in equipment and maintenance. The insertion of Artificial Intelligence and Machine Learning techniques adds sophisticated network control and monitoring resources.

- *Better frequency spectrum utilization.* With time and the replacement of 2G and 3G systems, the spectrum bands reserved for these cellular generations will naturally be incorporated into 4G and 5G systems and, in the future, probably be incorporated only by 5G and the next cellular generations.

Figure 11.2 shows shows how the spectrum should look in the USA when 2G and 3G are discontinued. It is interesting to note that above the 2.5 GHz frequency, the entire spectrum will be allocated to 5G.

Low frequencies (<1 GHz) can serve markets with lower user density and lower traffic volume. Frequencies above 1–2.6 GHz have another propagation

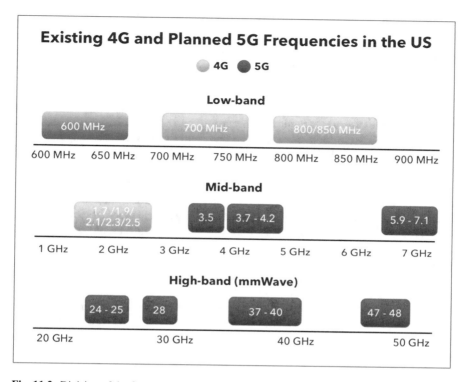

Fig. 11.2 Division of the frequency spectrum as it should look in the future for 4G and 5G in the USA according to operating frequency range, low, medium, and high. (Source: WAVEFORM)

characteristic and are intended for suburban and urban areas. From 3.5 GHz to 5 GHz, going up to 7 GHz, the propagation model changes and is intended to meet a higher traffic demand in very dense urban areas. When it comes to millimeter waves, in this case at 28 GHz, the use is completely different. This band is intended for very localized traffic, with tiny cells, practically hotspots, and is generally used to serve services that require high throughput.

The new services used in the 5G network work with large amounts of data. But don't forget that, despite improvements in modulation techniques, in order to achieve "broadband" it is necessary to use increasingly "larger spectrum bands," as Shannon predicted long ago. Since it is not always possible to use large continuous spectrum bands, aggregation techniques can achieve this result.

- *Evolution of the operational model of networks.* The physical characteristics of the frequencies that make up the spectrum are different when comparing high and low frequencies. The wave generated at low frequency has greater propagation capacity, i.e., the signal can reach further, allowing greater coverage area. The higher the frequency, the smaller its range. Consequently, the coverage area of the cell decreases, because high frequencies are easily blocked by trees, walls, mirrored glass, or any other type of obstacle. Today, much of the frequency spectrum available for mobile telephony transmission is already occupied, especially in the lower frequencies currently used by 2G, 3G, and 4G. This pushes 5G into the higher frequency bands of the spectrum, which implies working with sites with less coverage. In addition, 5G services require large bandwidth.

The graphic in Fig. 11.3 demonstrates the variation of the coverage area of cells according to the operating frequency. Regarding propagation characteristics, a cell operating at 700 MHz could cover an area that is 40 times wider than another operating at 28 GHz, considering for the latter a radius of 150 m, the typical coverage of this frequency range.

Figure 11.4 shows in detail how the spectrum should be used according to the needs of the market. It is clear that an operator should have available specific frequencies to be used according to the need imposed by the characteristic of the region where the service is provided. As explained, the ideal would be the sharing of all available spectrum by all operators, something perfectly possible with 5G technology.

So, to compensate coverage difficulties working with high frequencies, it is interesting to use different deployment strategies. This leads us naturally to the concept of heterogeneous networks, which combine macro cells, in which antennas operating at low frequency are used to cover large areas, with small cells, where higher frequency will be used to provide high data rates in a concentrated way, covering small areas, as shown in the picture.

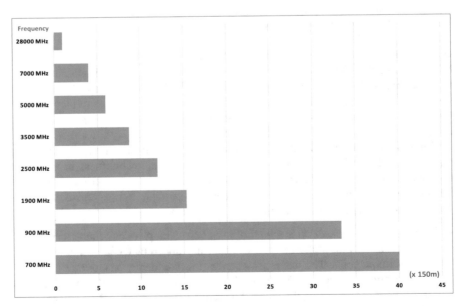

Fig. 11.3 Equivalent coverage area according to frequency compared to a 150 m radius cell operating at 28 GHz

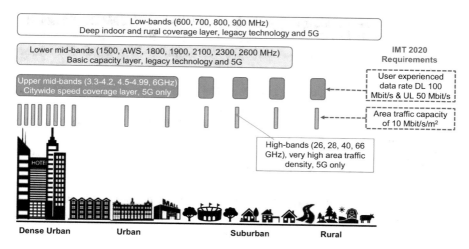

Fig. 11.4 5G spectrum to be used according to the characteristic of the market (Source: COLEAGO Consulting)

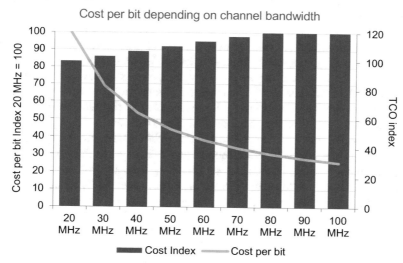

Fig. 11.5 Cost per bit depending on the available spectrum width (Source: COLEAGO Consulting)

The chart in Fig. 11.5 clearly demonstrates the need for broad spectrum for the provision of broadband services. Restricting or dividing the spectrum into small bands in order to foster competition makes deployment expensive and operational costs high and can backfire and make deployment unfeasible. The cost of building a network with 100 MHz of spectrum width is only about 20% higher than a network with 20 MHz. The cost/bit equals 30% of the cost of deploying a 20 MHz network when 100 MHz of spectrum width is available.

Those data are extracted from the excellent work done by COLEAGO Consulting: *"Estimating the mid-band spectrum needs in the 2025–2030-time frame"*. (https://www.coleago.com/app/uploads/2021/09/Estimating-Mid-Band-Spectrum-Needs.pdf)

- *Modernization of network architecture.* The availability of more abundant and competitive alternatives, both hardware and software, added to the need to meet the market in terms of traffic volume, almost instantaneous data processing, and a growing need for increased security, together with the projected increase of "things" connected, is leading to profound changes in the network. The insertion of functions such as Network Slicing and Edge Computing has led to a decentralization of the processing of network functions and a reorganization of resources in order to extract the best possible use of the system. Distributed Processing, which generates intelligence decentralization through the construction of mini and micro–Data Centers, allows the reduction of latency, the offer of higher quality services, distributes the traffic load through the network, and facilitates the insertion of new types of service. New technologies enable end-to-end communication monitoring and control, fast failure checking, zero-touch provisioning and maintenance, and powerful operator performance analysis and control tools.
- *Evolution of the networks business model.* But what about the promoters who make these innovations available to us, the cellular network operators? History has been hard to them; we are forced to acknowledge. In fact, they are the intermediaries of all the services we use, they are the ones that allow us to use their networks. Their main source of income, the "phone calls," especially the international ones, have practically ended, for the simple fact that telephone services no longer exist. They have become apps, with video calls and VoIP Apps, which can be downloaded for free and used without distinction. The only thing that operators still own exclusively is the numbering system of mobile phones, which stopped being phones, in the strict sense, a long time ago. Fixed telephony still exists, but it must be closed in a few years, as the holders of its concessions want. Fixed lines have been replaced by mobile lines, which are more flexible and easier to maintain. With fiber-optic networks, it has been possible to popularize content services (streaming, video, and TV) and offer connection, information, and entertainment at an affordable cost. But the telephone operators, who made this new world possible, are losing this huge and profitable line of business to the internet. This incredible mesh of physical connection called telephone network is being used as a means of distribution of data and content. It can include search engines like Google, music sites, YouTube, Netflix, and Prime Video, just to mention the most visited, not forgetting social media. Mobile networks allow access to all these wonders, a comfort few people can live without.

5G has the concept of business in its DNA, that is, it was designed as a service network and aims to offer excellence in terms of quality of service to the user. The FWA (Fixed Wireless Access) service, nicknamed "Fiber over the Air," proposes to offer 5G services to fixed end users, such as homes or businesses, and will become a major competitor to fiber optic last mile distribution, opening new monetization opportunities for operators. This implies a reduction in the use of fiber in fiber-to-the-home (FTTH) applications. But we must pay attention to the fact that the use of optical fiber should increase significantly in the internal network, either by interconnecting antennas of small cells or by extending the mesh of the transport network. 5G wireless is another large market that will allow even more services and quality to be added to mobile telephony. 5G will also enable the large-scale use of IoT services and applications for massive use, such as barcode readers, or critical use applications, such as remote surgeries.

At this point, someone might ask whether it makes sense for an operator to invest millions, or more accurately billions, to install a 5G network. Or how much of that investment only concerns spectrum usage rights. That question is worth many millions, and it is crucially important that it is answered before investments are made. For this, it is necessary to consider that 5G is not an isolated technology, but rather a technology developed to support technological innovations that should be applied in various IoT solutions, such as smart homes and cities, Industry 4.0, driverless cars, precision agriculture, medical applications, among others. These systems will generate a huge traffic demand that, in many cases, cannot be met by 4G. It will then be up to the operator to assess its investment capacity and the real needs of existing services in its area of operation to define how 5G will be deployed.

More industrialized countries have already begun installing and/or upgrading their networks to offer 5G services on a large scale, aiming to meet, for example, services such as 4 K/8 K UHD video, massive use of robotics in industry, applications of Augmented Reality, Virtual Reality, telemedicine, and massive control of IoT devices. The sale of 5G also makes operators more competitive in the cellular systems market, either by offering 5G-quality wireless telephony or by offering FWA services targeted at the residential user. The latest market innovation is the offering of private 5G, in which large users, such as companies or universities, acquire their own 5G system, which can even have operating frequency bands specially designated for private use.

Operators can take the opportunity to develop new revenue streams by exploiting Network Slicing, which allows them to provide differentiated services, serving each segment according to its specification (eMBB, URLLC, or mMTC).

Small Cells will be installed at key points in the network, providing access to a wide range of services. Indoor installations will also offer a rich market to explore, with infrastructure that can be built based on 5G or other protocols, such as Wi-Fi, which, by the way, has evolved considerably to meet the new needs of users. We have prepared a special chapter on Wi-Fi with more details on the subject. A wide range of companies related to telecom areas, such as installers, supply vendors, and backbone operators will benefit. In other words, all innovation is filled with uncertainties, but 5G certainly opens a huge range of opportunities for new ventures.

Final considerations In a way, the creation and popularization of new services follows the same line of evolution of data processing when PCs were launched, which enabled the popularization of access to information and has been providing immense opportunities to all social classes, especially the less favored. History repeats itself just a few decades later, when technological evolution opens the door to new services that will benefit society as a whole. Information processing, entertainment, the way we communicate, daily habits, commercial practices, medicine, industry and interactivity at all levels will be greatly benefited, generating immense opportunities for those who can take advantage of and ride the crest of this wave, which is expected to last for a long time. It is an interesting mental exercise to imagine the impact of all this!

Chapter 12
Digital Modulation in Detail

We have avoided going into unnecessary detail about how Digital Modulation happens, but the subject is so fascinating that we decided to create special chapter with detailed explanations for those who want to dig deeper into the subject.

We know that it is possible to change three characteristics of a carrier wave: its amplitude, its frequency, and its phase. But it would be interesting to correctly understand the meaning of this statement, and for that, we will go back to our high school days, when we learned trigonometry.

We know, since that time, that $cos(A)$ = adjacent side (ca)/hypotenuse (h), while $sin(A)$ = opposite side (co)/hypotenuse (h). To make it easier to understand, let's base ourselves on Fig. 12.1.

This means that, given an angle "A," the value of the cosine is equal to the projection of the hypotenuse on the horizontal axis. And the value of the sine is equal to the projection of the hypotenuse on the vertical axis. Great! Perfect! Got it!

What happens if the value of the angle starts to vary and, starting from 0°, goes up until it reaches 90°? Simple: the value of the cosine, which is equal to the size of the projection of the hypotenuse (let's admit that it is equal to 1) on the horizontal axis (ca - adjacent) decreases. When angle A is equal to 0°, the projection has the integral value equal to 1, but it goes decreasing until it reaches 0 when it reaches 90°. Meanwhile, the value of the sine varies inversely, corresponding to the projection of the hypotenuse (co - opposite) on the vertical axis. It starts with the value "0", for $A = 0°$, and grows until it reaches its maximum value, which in our case is equal to 1, when it reaches 90°. When "A" passes 90°, the cosine value grows again, but now with a negative value, while the sine value decreases until it returns to 0 when it reaches 180°. In this position, the value of the cosine passes through its maximum negative value, and so on. The figures obtained in this process are exactly the sine (in blue in the figure) and the cosine (in red).

Why are these curves so important? Most phenomena in nature vary *cyclically* and reach maximum and minimum values, according to a very well-defined rhythm,

© The Author(s), under exclusive license to Springer Nature Switzerland AG 2023
J. L. Frauendorf, É. Almeida de Souza, *The Architectural and Technological Revolution of 5G*, https://doi.org/10.1007/978-3-031-10650-7_12

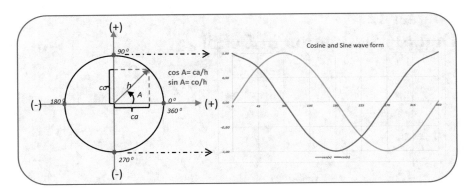

Fig. 12.1 Transforming sine and cosine into sine and cosine waves

like the seasons. Sines and cosines are offset by 90° and, for this reason, are said in *SQUARE or orthogonal*.

Sinusoids and cosines are fundamental to any propagation, be it propagation of sound, light, and, in our case, radio waves. The most traditional way to visualize a wave is when we throw a stone into a lake of completely still water. From the point where it falls, waves propagate in all directions, forming ups and downs that propel its movement until they dissipate, until they "die." This is exactly how sound, light, or radio waves propagate, with one very important difference: sound waves propagate in the air because they cause a disturbance in this medium like what happens to the stone when it falls into the lake. The luminous and radioelectric waves propagate even in a vacuum, like solar radiation or the signals we receive from satellites, radio, and TV stations.

We have learned how sine and cosine generate the sine and cosine waves, and that these are the waves we use. Let's see how we can take advantage of these waves to transmit information. We are interested in cellular systems, but the same reasoning is valid, for example, for the light signals that propagate in optical fibers and bring the internet signals to our homes.

In our analysis, we call the wave a "carrier," because it is the wave that will "carry" the information that interests us. For it to carry the information, we need to change it because it is exactly these changes that can be interpreted as information or a *bit*! We all use the terms "kilobits" (1,000), "megabits" (1,000,000), and "gigabits" (1,000,000,000) without knowing what exactly they represent. Let's make it clear what a bit is!

In order to send an information or bit, we need to change one of the wave characteristics, as shown in Fig. 12.2. If it is not changed, nothing is transmitted. What are these characteristics? They are the three already known characteristics: frequency (*F*), amplitude (*A*), and phase (*φ*).

What exactly is the *frequency*? We say that the sinusoid or cosine is formed by the variation of angle A and that, according to this cyclical variation, it reaches a maximum value, returns to zero, reaches a minimum value, returns to zero again, and continues in this rhythm indefinitely. The speed of this variation, this rhythm,

Fig. 12.2 Representation
of a wave

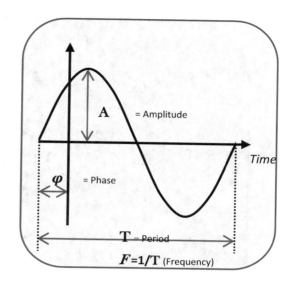

we call frequency; that is, how many cycles the sine or cosine wave completes per second. To this quantity, cycles per second, we give the name Hertz, abbreviated to *Hz* in honor of Heinrich Rudolf Hertz, a German physicist who discovered electromagnetic propagation. In our homes the electricity is supplied by the electric power companies as an alternating current (in a sinusoidal way), of 110 or 220 volts and of 60 Hz, that is, the electricity in our homes varies 60 cycles per second. In some countries, it is 50 Hz. This is one of the characteristics that we can change to send information. Every time a frequency is changed, that means I'm sending a bit of information.

Another characteristic of a sine or cosine wave is its *amplitude*, as we talked about above. The wave can be, for example, 110 volts or 220 volts. By changing the amplitude value, we are sending information.

The third quantity can be more difficult to understand because it is more subtle. It is the *phase of the wave*. If we look at Fig. 12.1, we notice that both the sine and cosine are almost identical in format. What makes them different is that one is offset from the other by 90°. It is like the sun and the moon – while one rises, the other dies, and vice versa. The difference between the two is exactly the value of the angle they form with respect to the origin. This angle is exactly the phase of the wave. When you change the phase of a sinusoid, you change its shape, and this change is interpreted as an information bit. Figure 12.3 summarizes what we mean. Any change in frequency, amplitude, or phase of a carrier wave is interpreted as an information bit.

Similarly, when a change is expected at a certain moment, and it doesn't happen, this is also information, in this case, that nothing has changed. In digital systems, we work with only two values, Zero (0) and One (1). Any letter (A....Z) or decimal number (1,2,3....,9,0) can be encoded in Zeros and Ones. For that to happen, we need several bits that represent each of the characters we want to encode, and that's

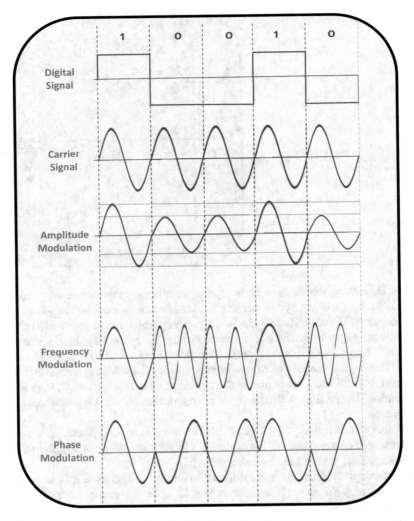

Fig. 12.3 Possible ways of modifying a Radioelectric Signal

why we created BYTE, which is nothing more than a set of 8 bits. Simple, isn't it? But some people confuse bits with bytes!

The carrier wave can be transmitted containing a multitude of information and, for this information to be decoded, it must follow a pattern. Therefore, it is important to define the *Time* or *Time Period,* or even better, *a Time Slot* for which it would be expected that some change occurs in the carrier signal. What and how many changes of the carrier are contained in this period? In Fig. 12.3, we may observe amplitude variation. During the Time Slot, two different levels of amplitude may happen, although, frequency and phase remain unchanged. In the case of frequency variation, the number of cycles varies, because that is exactly what transmits the information. In this case, two different frequencies may happen. Within the Time Slot, we may find just one cycle or two complete cycles. The last alternative is to change the phase of the carrier at the transition of a Time Slot. This "piece" of the carrier, which corresponds to the time that lasts the waveform that carries the information, we call *Symbol*. This time is very important, because it is what tells us how long the transmission of information lasts.

Let's complicate it just a bit more. In order to be able to assure the duration of a symbol, you must correctly identify the beginning and end of it, and for this, you need to *synchronize* the whole system, i.e., the *clock* must be the same from transmission to reception. One way to do this is to use in all cellular stations something we know well, the *GPS*. This is exactly who allows "set" all the clocks so that it is possible to identify when a symbol begins and ends.

Then someone may ask: is a symbol equal to a bit? The answer is NO, because a symbol can transmit a quantity of information higher than 1 bit, since a carrier may suffer more than a single variation. For example, it is possible to vary phase and amplitude simultaneously. The phase variation can happen in the transition from 0° to 180°, for example, as shown in Fig. 12.4. This is the simplest and most "robust" form of what we call *MODULATION*. Modulation is the name given to our ability to transform a sinusoid so that it is the "carrier" of the information we want to transmit. The North American Indians "modulated" the smoke, muffling or allowing the smoke to rise in order to transmit some code. Morse Code follows the same principle. This is how we can also do with a flashlight, turning it on and off. This is the act of MODULATING, causing some disturbance to be possible to transmit and interpret some code, which in our case are ZEROS and ONES.

BPSK (Binary Phase Shift Keying) modulation, shown in Fig. 12.4, is the simplest form, in which there are only two states, and therefore the *Symbol Rate* is equal to the *Bit* Rate.

For being the most robust form of modulation, it is employed in the transmission of synchronism information (for "setting" the clocks) and in sending pilots, that allows to inform the pertinent data to the transmission control. The circle, that

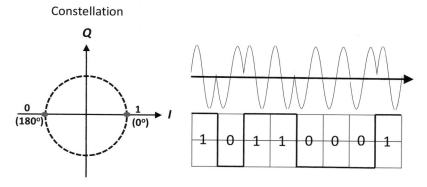

Fig. 12.4 BPSK – Binary Phase Shift Keying modulation

reminds us of trigonometry classes, shows the possible variations of the phases of the sinusoid. And, in the graphical representation shown in Fig. 12.5, we see the distribution of symbols, to which we give the name of *Constellation,* as if it were the celestial dome where all the "stars" are located. The symbols represent the possible positions of phase and amplitude of the carriers.

See that it is possible to transmit up to 4 bits of information in a constellation that holds 16 stars, that is, 16 different symbols generated by combinations of phase and amplitude of the carriers.

In the case of higher levels of modulation, the phase variation and, concomitantly, the amplitude variation are combined, keeping the frequency unchanged. This is the case of *QAM* (Quadrature and Amplitude Modulation) modulation, as shown in Fig. 12.5.

The most used modulation is the one that varies only the phase of the carrier signal, called *QPSK* (Quadrature Phase Shift Keying). In the case of QPSK, it is possible to transmit two bits within one symbol. In order to facilitate signal processing by electronic circuits, instead of adopting the digital system of 0 s and 1 s, it is easier to use 1 and $-$ 1, as shown in Fig. 12.6.

The block diagram in Fig. 12.7 shows, in a simplified way, how the modulation is done.

Let's analyze in detail how the modulation process happens according to Fig. 12.7. The bitstream, which contains the data set to be transmitted, arrives in serial format, one after the other. The first stage is to identify the period (time) duration of a symbol and check the number of bits that corresponds to the modulation used. QPSK allows the coding of 2 bits per symbol. If we observe Fig. 12.5, we see

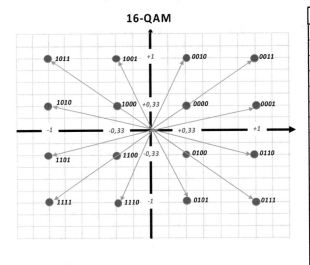

Amplitude (I) cos	Amplitude (Q) sin	Output Angle	Symbol
+0,33	+0,33	45	0000
+1	+0,33	23	0001
+0,33	+1	67	0010
+1	+1	45	0011
-0,33	+0,33	135	1000
-0,33	+1	113	1001
-1	+0,33	157	1010
-1	+1	135	1011
-0,33	-0,33	225	1100
-1	-0,33	203	1101
-0,33	-1	247	1110
-1	-1	225	1111
+0,33	-0,33	315	0100
+0,33	-1	293	0101
+1	-0,33	337	0110
-1	-1	315	0111

Fig. 12.5 16-QAM modulation

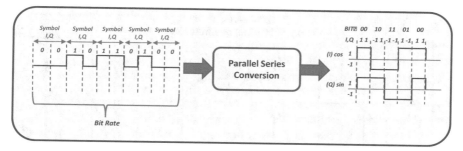

Fig. 12.6 QPSK two bits/symbol configuration

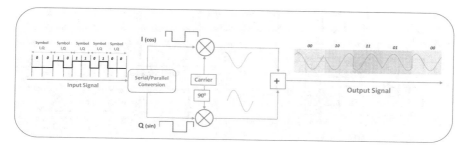

Fig. 12.7 QPSK (Quadrature Phase Shift Keying) modulation

that in 16-QAM the symbol is composed of four bits, because they combine, for each quadrant, three-phase variations, and two distinct amplitudes. In QPSK, the amplitude is always constant, and there are only four possible alternatives for the phases. This may be seen in the constellation diagram in Fig. 12.8.

Observing the pair of bits that corresponds to the signal symbol, the left one represents the value of I (cosine projection axis) in the vector diagram, and the right one, the value of Q (sine projection axis). These two combined values will then perform the corresponding phase changes of the cosine signal (I) and the sine signal (Q). Remembering, once again, that sine and cosine are signals offset from each other in "quadrature."

Notice that the generation of these two signals is obtained by an oscillator that generates the carrier of the signal to be modulated. The sinusoidal carrier is generated by displacing the phase of the cosine signal in 90°. The two resulting signals are added to generate the resulting output signal.

Figure 12.8 shows us the result of the sum of the two signals offset by 90° and how they end up getting the various shapes of the resulting waves, which symbolize each of the combinations of the corresponding bits.

It can be seen in the figure that the resulting signal is offset by 45° and that each resulting symbol can assume one of the four possible alternatives. These four alternatives are offset 90°, 180°, or 270° from each other, and can occupy, therefore, the positions of 45°, 135°, 225°, or 315° in the Constellation diagram. In the example

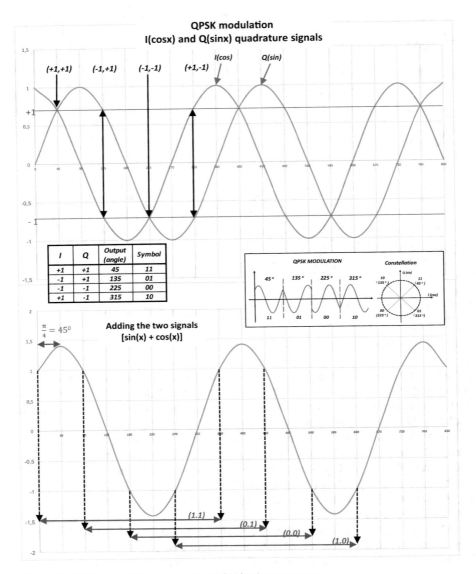

Fig. 12.8 QPSK modulation - I (cos x) and Q (sin x) components

presented above, each symbol is limited to a single cycle, but this is not what occurs in practice. Each symbol may contain more than one cycle, what matters are the transitions, initial and final, that characterize the symbol that represents the desired modulation scheme. This lag of 45° has nothing to do with a different kind of modulation, π/4 QPSK, which we will see next.

QPSK modulation has some drawbacks, as we will show. In many systems, it is preferred to use π/4 QPSK modulation. This is the case for cellular systems. The transitions that occur between symbols offset by 180° are inconvenient for electronic amplifiers because they require a very wide linear operating range.

Any transition caused by modulation generates signal scattering when looking at its frequency spectrum. Phase transitions when passing through the origin of the diagram result in a very sharp change of the signal shape in the time domain. Recalling what has already been said about Fourier Transform, which allows transforming events occurring in time into events occurring in the frequency domain, when a sinusoidal signal undergoes an abrupt transition, which is exactly what happens to the signal when it passes through the origin of the diagram, it generates in the frequency domain a very large number of harmonics of this signal. This greatly increases the number of components that make up this signal, as shown in Fig. 12.9, which greatly complicates the design of electronic circuits.

In order for the generated signal not to be distorted, electronic amplifiers must have very broad linear characteristics, which also allow the amplification of all components of the spectrum, which is not always possible. To avoid these transitions, we prefer to adopt π/4 QPSK, in which the transitions occur at 45° or 135° without crossing the origin of the constellation diagram, as shown in Fig. 12.10.

Someone might ask why this modulation is called π/4 QPSK and what is the difference to QPSK. What happens is that in π/4 QPSK eight positions of the constellation are adopted, being them: 0°, 45°, 90°, 135°, 180°, 225°, 270°, 315°, and 360° (=0⁰), that is, each position is 45° of its adjacent. To avoid the 180° transition, any of the transitions occur sequentially only between these positions, so that changes of ± 45° (in blue) or ± 135° (in red) are restricted, as shown in the constellation diagrams above. The goal is not to cross the center of the diagram. Any crossing of the diagram would imply a 180° transition, which is what happens in QPSK, and exactly what is undesirable.

Fig. 12.9 Frequency spectrum scattering due to the change of some parameter (phase, amplitude, or frequency)

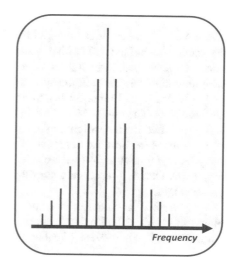

Frequency

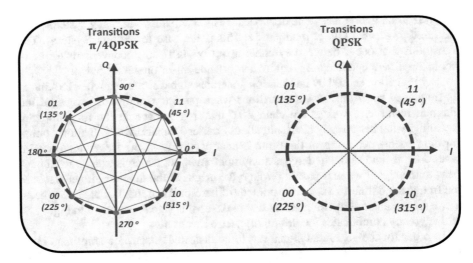

Fig. 12.10 Difference between π/4 QPSK and QPSK

In summary, QAM modulations are, in principle, like QPSK, in that more phases are used and variations in signal amplitudes are taken advantage of to extend coding levels and increase the efficiency of the spectrum used.

Was it too hard? We hope not. But, before we conclude this chapter, we would like to reinforce the understanding about OFDM(A) and SC-FDMA modulations, which are the fundamental basis of both 4G and 5G. For that, we decided to use a more playful way to facilitate understanding: telling a little story.

The Story of the Wireless G River Ferryman

There was a boatman on the bank of the Wireless G River who provided a crossing service for several users. The river was calm, but sometimes it got turbulent. So, the boatman developed a special technology to enable the cargoes to cross the river without much loss. He built the system he envisioned as shown in Fig. 12.11.

He basically had three types of customers, which he named QPSK, 16-QAM, and 64-QAM. The 64-QAM was the best of them, because he knew how to pack his cargo well, and the net weight of the cargo was equal to the gross weight. It was the one that made the best use of freight. The 16-QAM was also good, but it could only carry half of what the 64-QAM carried and used heavier packing. The worst customer was QPSK. It used very sturdy packaging, so it could only carry 1/3 of the gross weight.

They were all good customers, and to serve them, the boatman created two forms of transport. He called one direction of the crossing *Downstream* and created 12 lanes of transit. There were 12 cables, totally independent of each other. In case the

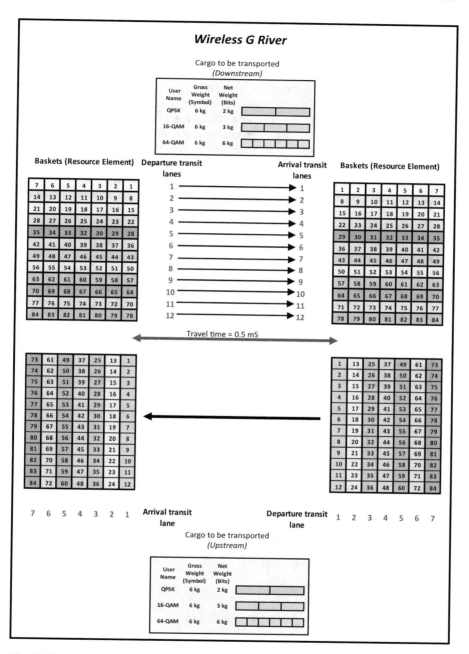

Fig. 12.11 Wireless G River crossing

river got turbulent, he didn't have to worry as he knew, it would only impact one of the lanes, not all, and only for a short time. On each cable, he carried baskets that he called *Resource Elements*, in which each part of the cargo of each client, called *Symbol* or Gross Weight, fit. Each set of seven baskets, which were transported in groups filling the 12 lanes he occupied, he called *Resource Blocks*, which contained 84 baskets (*Symbols*). The time for crossing the river last 0.5 ms.

In the beginning, each customer contracted the transport of seven baskets, which were dispatched one after the other. Sometimes, not all of them carried cargo. At that time, the boatman called this transport service *OFDM*. Soon he realized that he could use the empty baskets to carry loads for other customers, and he began to call this service *OFDMA*. Only that this system of 12 transport cables was very expensive, and he realized that customers sent much more load in the downstream direction than in the reverse direction. So, he decided to use a cheaper system with only one thicker cable that could carry 12 baskets at the same time. That system, he called *SC-FDMA*.

While the transport time in the downstream direction could last 0.5 ms, which made it slower but safe and reliable, in the reverse direction this time would have to be divided with the seven sets of 12 baskets, since he wanted to keep the same number of baskets in both directions and transport at the same time. This meant that transport in the opposite direction, which he called *Upstream*, had to take place in a much shorter and therefore less safe period. As the load was lower in that direction, it was worthwhile to maintain the scheme.

This is how the boatman managed to make all his customers very happy!

Hopefully, this story will help understanding the transport system used in both LTE and 5G.

Having fixed the concept that differentiates OFDMA from SC-TDMA, we can now focus on how the signal processing occurs in a Base Station and in the User Equipment, as shown in Fig. 12.12.

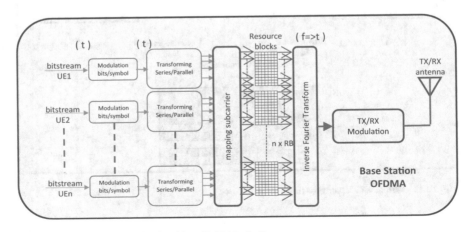

Fig. 12.12 OFDMA Base Station (downlink) block diagram

In the *downlink direction*, see that the base station receives messages from various users in the form of bit stream. Each bit stream is nothing more than a series of "0 s" and "1 s." These bits arrive separated in the *time domain (t)* – in series, therefore. They undergo in the first stage the digital modulation process, which in the case of LTE can occur in QPSK format, 16-QAM or 64-QAM. How the system discovers which is the format to be followed in modulation? Simple: by the input rate of the bit stream. At a high modulation level, the signal frequency is much higher than at lower levels, since this is a synchronous system in which everything occurs within very strict patterns of controlled time periods. Thus, the set of bits is transformed into symbols, which continue in a "serial" format, remaining in the *time (t) domain*.

The next stage is the transformation of the "serial" symbol sequence into a "parallel" format, where, sequentially, blocks of 12 symbols are "mapped" to modulate 12 subcarriers. The information is then subjected to an inverse Fourier transform to convert modulated subcarriers in the frequency domain into signals in the time domain. This process is called *inverse* Fourier transform, so called because normally the process proposed by Fourier decomposes a signal that occurs in the time domain into components in the frequency domain.

These subcarriers will undergo, in the next stage, an "up" conversion, that is, the set will modulate the RF carrier. In receiving the signal, that is, in the user's equipment, the reverse process occurs. The user equipment will always be demodulating the RF signals that are picked up by the antenna to know if the message is intended for it.

In the *uplink direction*, when sending messages from the user equipment to the Base Station, the modulation system is a little different, because it has one more stage than OFDMA. The reason to have two kinds of modulation is purely to save power and reduce the cost of user equipment. This is linked to the linearity of transmitters power stage. The explanation is simple.

OFDMA has a very large variation (excursion) in the power level required for signal transmission. In SC-FDMA this variation is smaller, requiring less power amplifier in the uplink. Let's examine how the system can join the contents of the uplink messages so that they occupy a set of 12 Resource Elements in a time period

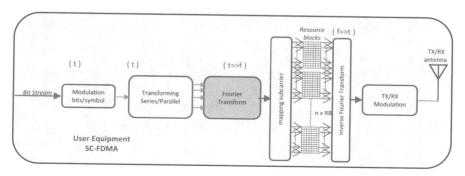

Fig. 12.13 Block Diagram User Equipment UE/SC-FDMA (uplink)

of only one Resource Element duration, and so that they are occupied only by content from a single user. Figure 12.13 helps to understand the explanation.

The message to be sent arrives at the transmission stage in the same way as downlink messages, i.e., in a bitstream of "0 s" and "1 s." They go through the modulation process, which converts bits into symbols, all occurring in the *time domain (t)*. In the stage of converting symbols from serial to parallel, the set of symbols begins to be prepared to allow conversion from the time domain *(t)* to the *frequency domain (f)*. In the next stage, a group of 12 "parallel," *time domain (t) symbols* are converted back to "serial," *frequency domain (f)* format so that they can be mapped into 12 Resource Elements that will modulate contiguous frequencies, *frequency domain (f)* therefore. This process is the Direct Fourier Transform. In the next step, in the same way, that occurs with OFDMA, the Inverse Fourier Transform takes care of so that subcarriers in the *frequency domain (f)* can be transformed again into a signal in the *time domain (t)*, completing the process. In the Base Station, where the reception of signals occurs, the process is reversed.

Chapter 13
IP Networks

Throughout all the chapters of this book, we talk about mobile telephony networks without going into details, beyond the specifics, and whenever necessary about the packet network or IP network. We thought it would be interesting to address this subject in a complementary chapter to provide the reader the opportunity to learn a little more about this very relevant topic in the universe of cellular systems.

Network Types

There are several types of communication networks, which can be identified by their own characteristics.

Personal Area Networks (PAN) are, in general, home networks in which a PC communicates with a printer, scanner, monitor, keyboard, mouse, etc.

Local Area Network (LAN) are private and corporate networks, to which multiple personal computers and servers are connected, most often using Ethernet protocol and routers, which allow access to public networks and the internet. The LAN is basically a computer network but may also include a digital telephone network. Yes, it is now possible to use digital telephone systems (VoIP), which eliminate the old PABX, typical of fixed-line telephone networks, widely used in offices, shops, and factories. In this case, the voice server of the packet network uses an element called a *gateway,* which interfaces with the public landline network.

Metropolitan Area Networks (MAN) are networks that serve users in a region – a city, for example. A MAN is formed by connecting several LANs, which are connected to a *backbone* (transport network) through high-speed connections. In addition, using a MAN, public telephone network operators can offer internet access via ADSL modems in addition to the fixed-line telephone service. Cable TV operators, on the other hand, provide voice services (VoIP) and high-speed internet access via DOCSIS modems in addition to the pay-TV service. The fiber optic networks make

J. L. Frauendorf, É. Almeida de Souza, *The Architectural and Technological Revolution of 5G*, https://doi.org/10.1007/978-3-031-10650-7_13

up the *backbone of* the most diverse operators, through which high-speed traffic passes. More recently, fiber has often been used as a distribution network (FFTH – Fiber To The Home), bringing broadband directly to consumers. Wireless networks, such as Wi-Fi, complement wired networks by enabling wireless connection of *endpoints*, such as computers, printers, and mobile devices.

Wide Area Networks (WAN) are the long-distance networks that interconnect all other networks.

An important and often mentioned characteristic is the network "synchronicity," which allows us to classify them as synchronous or asynchronous.

Synchronous networks are those that work so that everything happens according to a criterion commanded by time, in an organized, synchronized manner. Each element of the network (node) must act in a certain sequence of events and in a specific period of time.

Asynchronous networks are those in which each element is free to act as many times as necessary, checking only if the communication channel is free. Otherwise, it follows the criteria of waiting its turn, according to the priority and execution sequence. Asynchronous networks must have a message-sending format that identifies its beginning and end.

Network Topology

We will now see how they are organized, or rather, how is the topology of networks, how they are structured physically or even, how their elements are interconnected (nodes), and the logical structure, i.e., how information travels through them. They can be of various types:

Bus All elements are connected through a single conductive element, as if it were a bus. It is used when the network is local, of short extension, including internally in PCs. Each network component must wait its turn to communicate, avoiding collisions between data packets. The bus structure is the form adopted by the Service Based 5G SA (Stand Alone) architecture.

Ring This is still a bus, but with both ends connected. The ring structure is used in large extensions, such as fiber optic rings. The great advantage of this topology is that it allows the use of protection systems such as UPSR (Unidirectional Path Switched Ring), in which a redundant fiber ensures the transmission in case of rupture of a section, simply by reversing the direction of transmission of information, clockwise to counterclockwise or vice versa, so that the communication is reestablished.

Star Receives this name because it centralizes all communication between the elements through a main node. From this node, all information departs and arrives. In the specific case of cellular networks, network management is centralized at the

CORE, to which all network information, control data, and user data should converge to be processed. It is from the CORE that users of cellular networks have access to external networks and the internet. Remember that a BSC (Base Station Controller) (2G) can be connected to several BSs in Star topology and, likewise, an RNC (Radio Network Controller) (3G) can be connected to several NodeB according to the same topology. Another good example of this topology are Wi-Fi hotspots, in which a single device connects several users.

Mesh A configuration in which each element in a network communicates with several other elements in its own network. Some special wireless networks use this communication model. They are low mobility networks, called WMN (Wireless Mesh Networks). The disadvantage is that not only the traffic generated by the node, but also the traffic of nearby nodes may end up passing through this node, so that the total traffic tends to lose efficiency of the entire network.

Tree A star-shaped network in which there is a unit that concentrates traffic and connects to other similar networks in a branched format. Like this last one, there are several intercommunicating hybrid formats, of different topologies, in which each one has different characteristics from the others.

Communication Between the Elements (Nodes) of an IP Network

Once we know the topology of networks, let's understand how they communicate. In order to establish communication between the various elements of a network, rules are required, and these rules must follow a hierarchy so that each function is performed at an appropriate level. These levels are called layers. They can be organized in several ways.

ISO/OSI (International Organization of Standardization/Open System Interconnection) system Organized in seven layers, as shown in Fig. 13.1, each one is responsible for a very well-defined task, which makes these layers concerned with performing exclusively the function for which they were designed, within the established hierarchy. Thus, the lower layers provide services to the higher ones through a simplified communication interface. It is important to note that each layer has its own Header, a Header, which contains information pertinent to its hierarchical level. The data will only be available when layer 7 has been reached.

Layer 1 or Physical Layer Formed by bits, it is characterized by defining the physical aspects of the connection: mechanical construction, levels of electrical signals responsible for the connection, which may travel through metal cables, fiber optic cables, or even radio waves, which carry the bits generated by the various digital modulation systems. In addition to these parameters, transmission rates, transmission modes (duplex, half-duplex, or full-duplex), and network topology are also defined in the previously mentioned formats. At this level, in the case of 5G, there is still no differentiation of control data or data containing information. In systems in general, this level is responsible for important functions, such as signal processing, promoting the modulation of the radio frequency carrier to be used in transmission (up-converter), subcarrier modulation system (OFDMA, for example), and the use of advanced antenna systems (MIMO, massive MIMO and Beamforming). Ultimately, it takes care of the hardware processing of the transmission and reception system of the RAN (Radio Access Network) signals. It basically works with the binary transmission of bits, zeros, and ones (0,1).

Layer 2 or Link Layer (Data Link Layer/MAC – Media Access Control Layer) Its basic function is to "pack" the data to be transmitted, assembling the frames, which in turn will be delivered to the next layer. As described above, this layer takes care of the source and destination addresses and, especially, verifies the integrity of the received message thanks to the FCS (Frame Check Sequence) code, which is included in the frame construction. If the message arrives corrupted, it is requested to be resent in the procedure called HARQ (Hybrid Automatic Repeat Request). This procedure guarantees the quality of the transmission.

Layer 3 or Network Layer In general, this layer performs the function of packet routing, being responsible for managing the routing of data packets through the various nodes of the network to its final destination. It can perform several services:

• *Connection-oriented services*, in order to determine a channel or transmission medium between the two ends. In other words, it establishes a connection

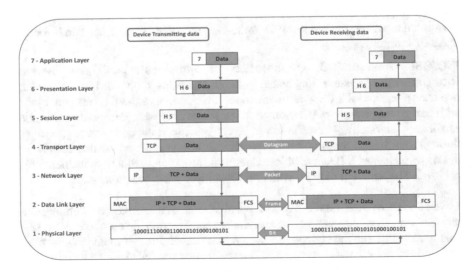

Fig. 13.1 OSI model communication structure

channel between the transmitter and the receiver. Once the connection has been created, all transmitted packets flow through a unique circuit (Bearer).

- *Not Connection-oriented services*, in which each packet is sent independently of the one sent previously, without establishing a fixed connection.

As each frame has a header, with information about the source and destination addresses, the various nodes of the network are responsible for identifying the best way to deliver the packets. Cellular networks also control message segmentation, adjusting the sequence in cases where there was a repetition of the transmission. In generations prior to 5G, this layer also processed the concatenation (ordering and sequencing) of the various messages in a way to make transmission more efficient, but at the expense of increased latency time.

Layer 4 or Transport Layer At the lower layers, transfers between nodes occur only on the network. The transport layer allows data to travel on a virtual circuit directly from source to destination, without concern for how data packets travel at the lower network layers. It is responsible for processing the end-to-end transfer from the source element to its destination. This occurs transparently, regardless of the technology, topology, or configuration of networks, without the occurrence of error and in the correct sequence (datagrams). This is possible because the lower layers have already taken care that the errors, which may have occurred, are corrected, allowing the sequence of packets to be correctly ordered. It happens that, as already discussed, to increase the efficiency of communication, several packets may have been sent in a non-sequential way, often due to multiplexing of messages originated by several users. It also makes sure that the message has been received by the destination user, without impacting the flow or generating overload in the system. In the 5G system, this layer manages encryption, header compression, and packet sequencing.

Layer 5 or Session Layer Designed to provide synchronization. Used in the processing of communication between network components, controlling the dialogue, and managing activities. When a connection is interrupted by a connection failure, the service is reestablished at the point where it was interrupted.

Layer 6 or Adaptation Layer (Presentation Layer) Takes care of the data formatting, its transformation, compression/decompression, and encryption of the content. The purpose of this layer is to convert the information received and intended for the application layer into an understandable format.

Layer 7 or Application Layer Takes care of the communication between applications, since each application has a specific protocol. These applications are like browsers, file transfers, email transfers, remote terminals, among many others.

Transmission Control Protocol/Internet Protocol (TCP/IP)

In addition to the OSI/ISO model, there is also the TCP/IP (Transmission Control Protocol/Internet Protocol) model, which is the language that computers in general use to access the internet. It consists of a set of protocols designed to allow a host to access the internet. TCP/IP contains four layers, which differ slightly from the OSI model, as shown in Fig. 13.2.

The TCP/IP Model is a simplified model of the OSI/ISO Model and consists of four layers:

1. *Network Access Layer or Link Layer*: Its goal is to enable the physical connection of the devices that participate in the network, and for this it uses specially defined protocols, such as PPP (Point-to-Point Protocol), Frame Relay, Ethernet, Token Ring, and ATM (Asynchronous Transfer Mode). In this physical layer, bits travel using various types of cables as a connection: twisted pair, coaxial, fiber optic, or even over the air, as is the case of cellular or Wi-Fi communication.

2. *Internet Layer*: Layer that allows devices connected to the network to send and receive packets. It uses the IP protocols (version 4 and version 6), like the OSI model layer. It is up to this layer to determine the ideal path to enable the delivery of IP packets to the right destination, promoting the correct routing within the path.

3. *Transport Layer*: Corresponds to the transport layer of the OSI model. It promotes the dialogue between the two devices, source, and destination, ensuring that messages arrive intact. There are two popular protocols used by this layer:

   - *TCP (Transmission Control Protocol)* is a more reliable protocol due to the number of checks, confirmations, and processing performed in order to ensure the delivery of packets in an integral manner to the end device. Due to all the verification process, it is a protocol with a high overhead.
   - *UDP (User Datagram Protocol)* is a protocol used in applications that do not depend on confirmations and checks for the integrity of the packets received. This procedure is left up to the application. Although less reliable than TCP, it is much faster and used in streaming audio and video applications.

4. *Application Layer*: presents the data to users in the appropriate format, according to the application used, in addition to controlling the dialogue and coding. The most used applications are:

   - *SMTP (Simple Mail Transfer Protocol)*, used to access e-mails.
   - *IMAP (Internet Message Access Protocol)* and *POP (Post Office Protocol)*, standard internet protocols for receiving email, which allow you to download messages from the server to a computer, smartphone, or tablet.
   - *HTTP (Hyper Text Transfer Protocol), a* protocol aimed at accessing the www (World Wide Web). It is used for the communication and transfer of

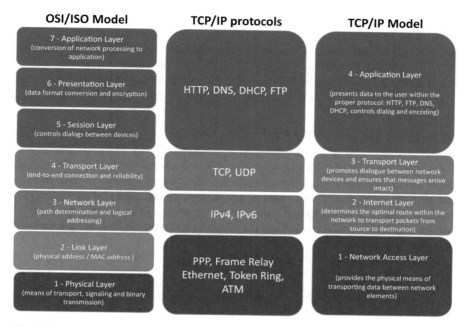

Fig. 13.2 ISO Model and TCP/IP Model

HTML documents (Hyper Text Markup Language), which occurs between a content server and the user.

- *FTP (File Transfer Protocol),* used to transfer files between a client and a file server. This protocol may require user authentication to allow access to the data. On public networks, authentication is not required. On private networks, the user must have a valid password.
- *DNS (Domain Name System), a* resource existing in networks that allows translating names, such as www.google.com, into an IP address and vice versa. It allows computers, services, servers, and other resources or elements available on the network to be identified and translates a name into an IP address, following a format that the devices that are part of the network can understand. The standard for an IP address (IPv4) is composed of four sets of numbers, each of which can range from 0 to 255, such as 185.201.0.25. The IP addressing range is from 0.0.0.0. to 255.255.255.255. With the scarcity of IPV4 addresses, the rapid growth of users and the addition of large number of devices, it was necessary to introduce IPV6, which has several advantages over the previous version, such as more efficient routing, better packet processing, direct data flow, simplified network configuration, support for new services and improved security. It has, however, the huge disadvantage of being a 128-bit hexadecimal address – something like 2830:0:1edf:a10f:b25c::5. It is very difficult for a person to memorize such a number, so the use of DNS server becomes imperative in this case. We also

note that technologies like IoT and 5G support IPV4 but are designed to operate with IPV6.

- *DHCP (Dynamic Host Configuration Protocol),* another feature that networks have for assigning a temporary address to a user so that the user can be identified while they are enjoying the services. Access providers have a range of addresses that can be assigned to their subscribers for as long as they are surfing the network.

We cannot fail to mention MPLS, which stands out as a protocol widely used in backbones and high-speed networks. To make access faster and more efficient, a technique called *MPLS (Multiprotocol Label Switching)* was developed, which allows the routing of packets traveling on the network with labels and enables the direction of a node to the next based on an optimized path, acting between Layers 2 (Internet) and 3 (Network) and avoiding the more complex process that usually occurs when following the usual routing procedure. For this reason, MPLS is called by some authors Layer 2.5.

Knowing the types of networks and how messages are processed, it is important to mention the types of messages that can travel on an IP network:

- *Unicast*: Point-to-point communication between two network clients or between a client and a server (information storage or processing device).
- *Multicast*: Communication between a network element, user or server, and a group of users or a group of servers.
- *Broadcast*: When the communication starts from a client or server, being directed to all the elements connected in that network.

How Information Is Transmitted on the Ethernet Network

In a network, the information flows are organized in blocks called "data packets," and these packets occupy their places in a kind of envelope, the frame. According to the IEEE802.3 format, the Ethernet frame is composed of a preamble, which identifies its beginning, followed by a kind of header, where the destination and origin addresses (*MAC* Address – Media Access Control Address) are indicated, as well as the type of message being transmitted. Finally, the field containing the information (encapsulated data) appears. At the end of the frame, there is an error identification code, FCS (Frame Check Sequence), aimed at verifying the integrity of the transmitted data, as shown in Fig. 13.3.

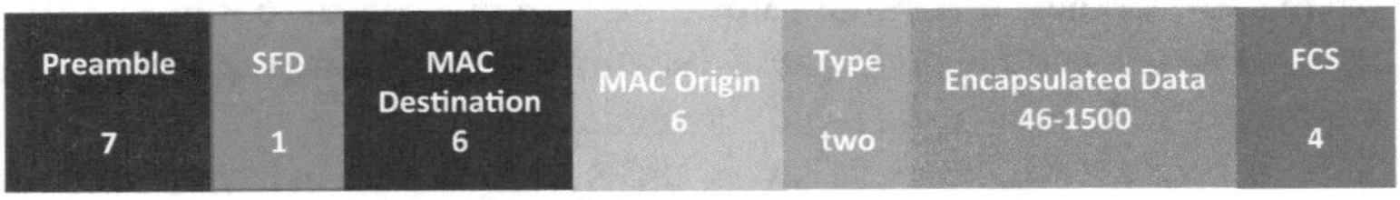

Preamble	SFD	MAC Destination	MAC Origin	Type	Encapsulated Data 46-1500	FCS
7	1	6	6	two		4

Fig. 13.3 Ethernet frame structure

Data Center and Servers

It is important to characterize the role of the *Server* in a network. With the growing demand for data processing and storage capacity, the concept of having processors and file storage elements in special places, known as Data Centers, emerged. These are generally facilities with adequate infrastructure and capacity to operate 24 hours/day, 365 days/year, with complete security and reliability, with climatized environments and continuous energy supply, without any possibility of interruption. These Data Centers can be public, shared, or for exclusive use, such as banks and large organizations. But nothing prevents the existence of home servers or used in small businesses. What characterizes a server is the fact that it has an IP address that allows access by authorized users. Within this concept, there may even be servers in home networks or small businesses that have their own network. There are, in fact, several types of servers: print, email, local file, cloud file, WEB (HTML), in addition to the aforementioned DNS and DHCP. With the increasing use of virtualization in telecommunications networks, servers have come to be used in many different types of applications, often replacing even expensive systems of dedicated hardware. A simple example that is being widely used is the voice server (VoIP), which offers services like a digital/IP PBX, allowing conventional landline phones to be connected to a digital network through a small adapter, which converts analog signals into digital signals. The gateway promotes the interface between the local network server and the fixed-line operator, which already receives the traffic in digital format. The great advantage is the elimination of the archaic PABX.

There is a special type of network that should not be forgotten: *VPNs* (Virtual Private Networks), private networks with restricted access. With a VPN, you have an exclusive channel within a public communication network, in which data passes in encrypted form. This allows physically distant users to use the internet to securely connect to the local network using a remote access client software. VPN uses special centralized management software to enable functions such as authenticating remote users, hiding the IP address of your users and ensuring QoS. In any online browsing within a VPN, the data exchanged in messages are protected and cannot be accessed by the ISP (Internet Service Provider), much less by third parties.

Network Elements

We talk a lot about *Network Elements* without mentioning what they are or their functions. In fact, Network Elements are all the devices necessary for the operation of networks. Let's get to know them better.

Repeaters Devices that serve only to amplify signals that are normally attenuated by the physical means of transmission (cable or wireless).

HUB Concentrator of connections, a kind of multiconnection adapter used in electrical outlets, which allows the connection of everything to everything. It is a device devoid of intelligence, which makes information available to all connected users.

Manageable Switch or Concentrator Intelligent hub can handle multiple simultaneous messages. The information is forwarded only to the recipient and cannot be accessed by other users connected to the concentrator. Its capacity is limited because it does not optimize routes between switches connected to the same network.

Bridges Similar to the switch, with the same disadvantages and even fewer ports. They can, however, connect networks that operate with different operating systems, such as Linux and Windows, serving as an interface and enabling communication between them and sharing resources.

Router Equipment with more intelligence, whose algorithms allow to treat information in order to optimize the route to the destination of the message. It may contain a Firewall, protection software against hacker invasion. It may also be associated with switches to increase its capacity or be connected in cascade with other routers. In this case, there will always be a master router in charge, while the others act as slaves, which obey the command of the master unit.

Gateway Device that allows the interconnection of distinct networks, while all other devices listed above participate exclusively in the same network. A gateway can be a media converter, for example, acting as an interface between a fiber-optic network and the home Wi-Fi network. But it can have more complex functions, as is the case with cellular networks, where the gateway acts as a protocol converter. Eventually, the IP address that a user uses on a particular network is for exclusive use within that network. When he intends to travel on another network, he needs a valid address on this external network and an element that mediates this intermediation. This is exactly the role of the gateway. By working in an intermediate zone between networks, the gateway is likely to be more susceptible to attack. For this reason, it may be associated with security features such as a *Proxy Server*. All requests directed to the internet pass through the Proxy, which analyzes the content of the packets and, after security checks, passes them on to the internet, protecting the user's data while hiding the network addresses. The same process occurs with the reply to messages. The gateway may also be associated with a Firewall, to ensure the security of communications.

Transmission Media Used in the Packet Network

In order to complete this chapter, it remains to discuss the transmission media used to build a network, which can be either "wired" or "wireless."

Wired Network

The best known "wired" physical networks are of the coaxial type, which was the first cable used by networks and works in the bus topology, in which everyone is connected by several cables to the same physical bus through BNC-type connectors. They allow long distances, but the maximum speed is 10 Mbps. The most used cables today are the twisted pair type, also known as *Ethernet Cable*, which use *RJ-45* connectors. There are several categories of cables, whose maximum transmission speed depends on the maximum distance used. Figure 13.4 shows the speed x distance differences according to the category.

Fiber Optic Networks

In addition to these two physical networks, the most important ones for overcoming long distances are fiber-optic networks, which use light beams as the carrier wave. The advantage is that they are immune to human-generated noise, and the signal can be regenerated when necessary, allowing transmission over long distances. They are usually cables with several fibers that can be buried or installed on poles. There are two types of optical fibers, Single-Mode (SM) and Multi-Mode (MM). Figure 13.5 illustrates the differences.

In SM fiber, propagation occurs via a single light beam, while in MM fiber, light propagates through *multiple paths (modes)*, which means that more than one light beam can be sent at the same time due to the different angles of incidence.

Single-mode fibers can span great distances, including up to 80 km. Multimode fibers reach a maximum of 300 m. These are reference values and may change depending on the transmission speed and technology used.

One can convert multiple signals to wavelengths slightly different from each other, multiplexing those wavelengths at the source, transmit them through fiber optics, and demultiplex those wavelengths, separating and recovering each of the signals at the destination. This technique increases the transmission capacity of the

Category	Maximum Speed	Maximum Distance (m)
Cat5	100 Mbps	100
Cat5e	1 Gbps	100
Cat6	1 Gbps	100
Cat6a	10 Gbps	100
Cat7	10 Gbps	100
Cat8	40 Gbps	35

Fig. 13.4 RJ-45 Ethernet cable types

Fig. 13.5 Types of optical fibers

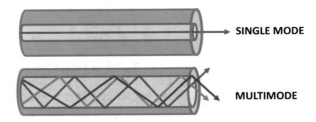

SINGLE MODE

MULTIMODE

system. Wavelength Division Multiplexing (*WDM*) in its variants *CWDM* (Coarse Wavelength Division Multiplexing) or *DWDM* (Dense Wavelength Division Multiplexing) allows better use of existing fibers, without the need to lay new cables.

Wireless Networks

Basically, the two networks that deserve our attention are cellular networks and Wi-Fi. As this book is entirely devoted to the former, it is now up to us to talk about the latter.

Used to provide wireless communication indoors and outdoors also, with short-range coverage, Wi-Fi public wireless networks are widely known and present in coffee shops, malls, squares, schools, and several other places, allowing us to browse without difficulty, use social media and check emails. However, Wi-Fi still has security vulnerabilities and often poor transmission quality. Even with these problems, Wi-Fi networks are most likely among the most used networks in the world, because virtually all devices that allow communication are equipped with a Wi-Fi interface. The biggest advantage of this technology is that it operates in spectrum bands that do not require a license to use, which has allowed the assembly of small networks in homes, commercial, and industrial establishments.

ISPs (Internet Service Providers) took advantage of this flexibility to launch their first Wi-Fi access systems using outdoor devices mounted on top of buildings, creating large hotspots, and covering most small locations. These systems have gradually been replaced by fiber optics. It is an invaluable service, which has allowed them to bring Wi-Fi access to regions overlooked by large operators.

Wi-Fi was launched in 1997. From launch to 2020, its evolution has been exponential, like that of cellular technology. In just over 20 years, its transmission rate capacity has multiplied by 5000, from 2 Mbps to 10,000 Mbps. These values are theoretical, only achieved in the best conditions, as we will see below. Since its launch, Wi-Fi has gone through six distinct generations, which are shown in Table 13.1.

A lot happened in this period. The first fact was the release of unlicensed spectrum. It started with the 2.4 GHz band, also used by cordless phones, Bluetooth, microwave ovens, and some other devices, which can interfere with transmission, reducing operating efficiency. Because it is already a high frequency and cannot

Table 13.1 Wi-Fi evolution

Generation	WiFi	WiFi 1	WiFi 2	WiFi 3	WiFi 4	WiFi 5	WiFi 6	WiFi 6E
IEEE	802.11	802.11b	802.11a	802.11g	802.11n	802.11ac	802.11ax	802.11ax
Release year	1997	1999	1999	2003	2009	2014	2019	2020
Operating frequency (GHz)	2.4	2.4	5	2.4	2.4/5	5	2.4/5	6
Inner/outer coverage radius (m)	20/100	35/140	35/120	38/140	70/250	35	?	?
Available spectrum width (MHz)	80	80	180	80	80/180	180	80/180	1.200
Channel width (MHz)	20	20	20	20	20/40	20/40/80/160	20/40/80/160	20/40/80/160
Maximum transmission rate (Mbps)	2 Mbps	11 Mbps	54 Mbps	54 Mbps	600 Mbps	6.8 Gbps	10 Gbps	10 Gbps
Spectral efficiency (bps/Hz)	0,1	0,55	2,7	2,7	15	42,5	62,5	62,5
Maximum number of single user streams	1	1	1	1	4	8	8	8
Maximum number of multiple user streams	AT	AT	AT	AT	AT	4 (DL only)	8 (UL/DL)	8 (UL/DL)
Modulation system	DSSS, FHSS	DSSS,CCK	OFDM	OFDM	OFDM	OFDM	OFDM/OFDMA	OFDMA
Maximum modulation level	DQPSK	CCK	64-QAM	64-QAM	64-QAM	256-QAM	1024-QAM	1024-QAM
Number of OFDM/OFDMA subcarriers	AT	AT	64	64	128	512	2048	2048
Subcarriers spacing (kHz)	AT	AT	312.5	312.5	312.5	312.5	78.125	78.125

operate at high power levels, which could harm the health of users, its range is limited. The first generation operating in this band had a range of 20 m indoors and 100 m outdoors (line of sight). Subsequent generations were able to extend this range to a maximum of 38 m for indoor use, and up to 140 m when used outdoors in open space. Only Wi-Fi 4/IEEE 802.11n was able to overcome this barrier, allowing it to operate at distances of 70 m indoors and 250 m outdoors, using the 2.4 GHz band.

Wi-Fi was designed to be a hotspot, that is, to serve a small area. Coverage has never been the most important parameter, but throughput yes. In the latest versions of Wi-Fi, the maximum throughput has grown substantially, from 600 Mbps in Wi-Fi 4 to 6.8 Gbps in Wi-Fi 5, reaching 10 Gbps in Wi-Fi 6. Another major concern is the number of users that can share the services provided, because the demand for more access has grown exponentially, as we well know.

With the saturation of the 2.4 GHz band, the 5 GHz spectrum was released, and more recently 6 GHz (Wi-Fi 6E). As always, the higher the bandwidth requirement, the higher the frequency. While the 2.4 GHz spectrum is 80 MHz (4 channels of 20 MHz), the 5 GHz spectrum is 180 MHz and now, with Wi-Fi 6E, the available bandwidth is 1200 MHz. In addition, the width of each channel has also evolved, and the original 20 MHz channel made available for the first three generations of Wi-Fi now supports a 40 MHz band in Wi-Fi 4, and from Wi-Fi 5 onwards, channel sizes can vary from 20, 40, 80, or 160 MHz at the 5 GHz frequency. But, if the frequency goes up, the signal attenuation also goes up too! In order to maintain the coverage area, more efficient modulation systems had to be sought.

In this aspect, the parallelism with cellular technology is perfect. The first generations used DSSS/FHSS (Direct Sequence Spread Spectrum/Frequency Hopping Spread Spectrum) and DSSS/CCK (Direct Sequence Spread Spectrum/Complementary Code Keying) as modulation systems, modulation formats like those used by the first generations of cellular systems. This period was very short, just 2 years, and Wi-Fi soon innovated, moving far ahead of cellular technology by adopting OFDM as early as 1999, something that would happen with 4G in 2010.

In addition to adopting more efficient modulation systems, Wi-Fi, like cellular, had to raise the modulation levels, evolving from 64-QAM in 1999 to 1024-QAM as of 2019, a brutal leap in terms of technology. With the new version of Wi-Fi6E or IEEE 802.11ax, Wi-Fi6 adopts, like 5G, OFDMA modulation and uses MU-MIMO 8×8 (Multiuser Multiple-Input Multiple-Output) antennas, which allows it to connect up to four times more devices than previous standards.

Increased speed and capacity are essential because the volume of mobile data traffic is likely to grow rapidly in the coming years. Wi-Fi must meet the demand for new technologies and work with next-generation applications such as 4K/8K HD streaming, Augmented Reality (AR) and Virtual Reality (VR) video, and IoT, and support higher-capacity devices in high-density environments such as university lecture halls, malls, stadiums, and factories.

With so much similarity between Wi-Fi6/Wi-Fi6E and 5G technologies, a question arises: will they coexist together or will 5G take the place of Wi-Fi? First of all, we must bear in mind that both technologies are new, so devices to operate with 5G or with Wi-Fi6/Wi-Fi6E require the user to have specific interfaces to access the services. This means changing equipment or investing in external devices that enable these new technologies to be used.

Overall, both Wi-Fi and 5G are critical to the future of wireless networking, but Wi-Fi will remain a preferred access medium for use in homes and businesses, especially for indoor networks. When considering deployment, maintenance, and scaling costs, Wi-Fi is a very reasonable choice and is ideal for indoor wireless connectivity. However, for medium or large enterprises that can invest a bit more in their system, Private 5G can be an interesting alternative if you consider the services (internal and external) enabled. Applications such as smart buildings and industrial IoT need varying degrees of network connection sophistication, which may require the deployment of small 5G cells. Sectors such as hospitals, retail, and education will be able to assess their real need and choose between Wi-Fi and 5G options.

On the other hand, 5G can be designated for certain uses, such as communication on a bullet train at 200 miles per hour, internet and app usage in a moving car, mission critical IoT device usage in factory process automation, healthcare, energy, and many other sectors. 5G mobile is likely to be the preferred method in outdoor networks.

Wi-Fi can only operate in an unlicensed spectrum, which has favored its spread so far, but 5G can operate in both licensed and unlicensed spectrum. In addition, if access to the internet network always depends on an operator, it is more advantageous to install a 5G Femtocell than a Wi-Fi6/Wi-Fi6E, since the dialogue between cells follows the same protocol as the cellular network, which is different in the case of Wi-Fi. Everything will depend on the cost and the volume of available units. We can assume, however, without any specific knowledge of cause, that it would not be a surprise to learn that, in a short time, the manufacturers of the two technologies, cellular and Wi-Fi, manage to "package" both in a single chip. Let's note that the cellular/Wi-Fi handover is no longer a problem since 4G.

Of course, when you consider things like traffic volume, number of connected units, security, and latency, 5G naturally stands out. Manufacturers have been working hard to improve more fragile aspects of Wi-Fi, such as security. This aspect has been considered in Wi-Fi6, which offers increased security with Wi-Fi Protected Access 3 (WPA3) and interference reduction, providing a better quality of experience for the user, which is especially useful for securing enterprise Wi-Fi networks.

We live in a time when technological miracles are announced almost daily. In any case, for now, everyone advocates the peaceful coexistence between the two technologies. Only the future will tell!

Chapter 14
Artificial Intelligence (AI) and Machine Learning (ML)

In many chapters of this book, we have mentioned the terms *Artificial Intelligence (AI)* and *Machine Learning (ML)* without explaining exactly what these technologies mean. It is not our intention to exhaust the subject, quite the contrary. We intend, in the following paragraphs, only to give an idea of the concepts behind these expressions. Often, when we hear these names, we have the feeling we are addressing something that, according to the most extreme theories, could endanger the existence of human beings.

These technologies are really disturbing. But we must clarify that they are based on logical and mathematical principles that are easy to understand and that have their own limitations. The most important thing, as already occurred when the computer was created, is that man emulates in practice the functioning of his own body. This is the case of von Neumann's architecture, used in the computer in which this text is being written and that is shown in Fig. 14.1.

The concept is quite simple: there is a *Data Input Unit* that provides the information, the *inputs*. The *Control Unit* tells the *Logical and Arithmetical Processing Unit* what to do with the information. There is the *Memory Unit*, which stores all the received information as well as the processed information. The Control Unit is programmed to know *what, when, and how it* should manage the processing of the received information. The input, after being correctly stored, may require some logical or arithmetic processing to be made available externally through the *Data Output Unit*, the *output*.

It is interesting to note that processing is always and only logical or arithmetic, and that any more complex process must be carried out by means of an *algorithm*. An algorithm is nothing more than a sequence of instructions or operations (logical and/or arithmetic) to be executed by the Logical and Arithmetic Processing Unit, under the command of the Control Unit according to a routine that translates a reasoning. This reasoning, expressed in the form of instructions (programming), is something to be executed in a systematic way in order to achieve a certain result. The computer is equipped with an Operating System (Windows, Linux, etc.), which

© The Author(s), under exclusive license to Springer Nature Switzerland AG 2023
J. L. Frauendorf, É. Almeida de Souza, *The Architectural and Technological Revolution of 5G*, https://doi.org/10.1007/978-3-031-10650-7_14

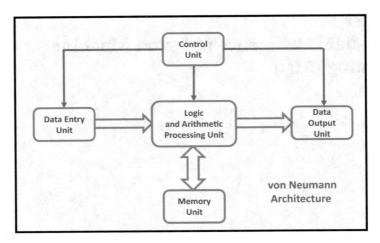

Fig. 14.1 von Neumann architecture used in computers

is the communication interface with its users and allows running specific programs for processing applications (Excel, Word, PowerPoint, etc.). The Operating System is a software that supports the basic functions related to the management of hardware resources, file systems, and programs on a computer, such as sequencing of tasks, running applications, and control of peripherals, while applications are software that work as a set of tools specially designed to perform specific tasks and work on the user's computer.

Artificial Intelligence is based on these same principles. According to the father of AI, John McCarthy, *"Artificial Intelligence is the science, the engineering, that allows simple machines to be turned into intelligent machines by means of ingenious programs developed for computers."* Ultimately, it is the art of crafting software that acts intelligently. This is all done through studies that evaluate how the human brain thinks, how it learns, decides, and works while solving a given problem and consists of applying the result of these studies in the generation of intelligent software and systems.

The goal is to create systems that exhibit "intelligent behavior," such as the ability to learn, demonstrate, explain, and recommend solutions to users. Its applications cover diverse areas: biology, psychology, philosophy, sociology, linguistics, neuroscience, mathematics, and engineering.

What would be the basic principle of how IA works?

- In the real world, we are inundated with a huge volume of data and information.
- These data and information do not come to us in an organized way, but randomly.
- Data and information are constantly being updated.

What IA primarily does is handle data and information in an efficient and organized manner such that:

- The transfer of data is perceptible to the party providing the data and information.

- Be able to update the data and correct the errors.
- The data and information can be used in various situations, even if it is incomplete or not completely correct.

AI requires that data and information can be processed very quickly due to the complexity required by the programs used. Obviously, this has only become possible with the enormous advancement of the processors employed.

We can cite some concrete examples of the use of IA:

- Games, like chess. In 1997, IBM's Deep Blue Chess Program defeated world champion, Garry Kasparov.
- It is already possible to do simultaneous translation in several languages using programs installed even in mobile phones and tablets. For example, this book was originally in Portuguese and translated to English using a translation program called Deepl. There are also programs for voice and sound recognition.
- Visualization systems are used in face recognition, retina recognition, geographic area recognition with satellite visualization, disease diagnosis by image evaluation, and even signature recognition.
- Robots perform tasks based on programming from the most diverse inputs provided by sensors for temperature, pressure, heat, motion, sound, shock, etc. They can learn based on mistakes made and can adapt to working conditions.

Since AI is based on human factors, it is worth enumerating the *Intelligence Types*:

- Oral Intelligence
- Musical Intelligence
- Logical and Mathematical Intelligence
- Spatial Intelligence
- Body Intelligence
- Intrapersonal Intelligence
- Interpersonal Intelligence

What are the *Components of Intelligence*?

- *Reasoning*, which can be intuitive or deductive. Many advocate that even reasoning that we think is intuitive is often actually deductive, as it is based on data accumulated over years of experience and occurs without logical reasoning being explicit. This is the case of the advice of the elderly, adopted in several ethnic groups. This concept has a lot to do with AI.
- *Learning*, which can occur in a variety of ways:

  - Through *hearing*, characterized by the habit of listening and repeating to facilitate memorization.
  - Though *observing*, which occurs after being exposed to, or participation in something that happened.
  - Through *motion*, which can be acquired with repetitive physical exercise, allowing progressive improvement.
  - Through *repetition*, which stems from observing and imitating someone.

- Through *perception*, which can be realized based on facts and events that happened in the past.
- Through *comparison*, which involves comparing a new event with something based on previous experiences.

- *Solving Problems*, a process in which a problem is proposed, and one evaluates all alternatives until a satisfactory result can be reached. It involves choosing the best alternative among those available.
- *Perception*, that involves being exposed to a situation, understanding the context, and processing information based on previous experience.
- *Linguistic*, which refers to language-related skills such as listening, understanding, speaking, or translating in written or colloquial language. What is important is the interpretation of the information received.

What would be the main difference between Human and Machine Intelligence?

- People make decisions based on patterns, while machines' outputs are based on available data and rules.
- People accumulate information and use them accordingly, while machines process decisions based on algorithms.
- People can perceive an object, even if a part of it is missing. Machines cannot.

Let's Try to Understand How AI Works?

First, let's look at something called *Fuzzy Logic (FL)*. Fuzzy Logic is a problem-solving method that resembles human logic. It mimics the logic we employ in situations that cannot be characterized as YES or NO, RIGHT or WRONG, TRUE or FALSE. There can be numerous ranges between the extremes so that it is possible to narrow the imponderability of the result. The advantages of the implementation of Fuzzy Logic are several: the concept and implementation are quite simple and can be modified only by adding or subtracting intermediate ranges; is an algorithm simple to build and to understand; and can be the solution for various types of applications, including medicine, because it is a model that replicates the decision-making of human beings. The disadvantages are as follows: it can only be applied to problems that do not require great precision; the logical structure is only easily understood when it is simple and there is no systematic procedure to be implemented.

There is another system employed by AI called *Expert Systems (ES)*. These systems exhibit high performance, can be understood easily, and are highly reliable and very sensitive. On the other hand, they serve only as a basis for decision making, without the condition to fully replace the human being for not having the capacity of understanding inherent in it, which is to produce appropriate responses which require interpretations. These systems depend on a very robust knowledge bank, based on existing data, information, and accumulated experiences. Data entered into the database of an expert system are added by experts in each domain, and the query, in general, is made by non-expert users who wish to acquire information. This system is totally based on a decision tree of the IF-THEN-ELSE type.

Machine Learning, in a simplified explanation, is the way a computer learns things without being programmed <u>specifically</u> to do so. That's not to say that it hasn't been programmed to perform functions. Yes, it has been programmed, but not for a particular function. While we humans use our own experiences to learn, a machine learns from logical instructions created to do so. Programs are created from a set of rules or algorithms that a computer can execute. From data and from these algorithms, a machine can "think" (run programs) and give answers.

The machine's response is predictions or "clustering," i.e., it can determine the existence of patterns by grouping data according to similar characteristics.

Prediction is a process in which, based on input data (variables), it is possible to predict the result of the output after processing. This can be used when the data can be precisely identified between the inputs and the outputs. This procedure is called *Supervised Learning*. It might be easy to predict a price of a product based on existing data of the products' attributes.

Problems involving predictions can be divided into two categories:

- Problems that can be solved by means of *Prediction by Regression*. Regression is a statistical methodology in which the result is a real numeric value that expresses a certain quantity and that can be estimated from a mathematical formula, determined by the analysis of a considerable volume of variables. For example, you can estimate how long a trip will take based on the route, traffic conditions, and road surface.
- Another alternative is focused on *Prediction by Classification* according to a certain category, in which the parameter evaluated is purely a YES or NO decision. That is, it fits or not in the chosen category.

Thus, a prediction is quantitative when it uses regression and qualitative when it does or does not fit a particular category.

The most used algorithms are linear model, branched model, and the neural network model.

It is important to know the terminology used in Machine Learning. For this reason, we will clarify some terms that are used:

- *MODEL*, that represents a process according to mathematical equations. It might be possible to predict some event based on a model once we know the variables.
- *TRAINING* is used to create a model based on accumulated information. It is used to "teach" a machine.
- *TRAINING SET* is the reliable set of data used to develop an algorithm.
- *CLASSIFICATION* is used to categorize a set of data according to some attributes or variables.
- *FEATURE* exactly the attribute that will be observed (Independent Variables), stored, and applied to the model to obtain the result.
- *ALGORITHM* as already clarified is the set of mathematical rules used for solving a problem.

- *REGRESSION* is the methodology used to define a function that represents the behavior of a certain amount of data. Thanks to the knowledge of this mathematical function, it is possible to predict the behavior of some parameters being evaluated.
- *TARGET* is the value that will be found when applying a given model. In statistics, it is called dependent variable.
- *TEST SET* is the amount of independent data (not to confuse with Training Data) that may be used to evaluate the performance of a model.
- *OVERFITTING* reflects the situation where the prediction model employed in the analysis of the data provided is too complex. The resulting predictions do not reflect the fundamental reality of the relationships between the features and the target.

The main algorithms used in ML are:

- *LINEAR MODEL* which uses an uncomplicated formula to search for the best answer following a line of reasoning created from a set of data. It is good because it is simple. The variable whose prediction is desired (dependent variable) is represented by an equation involving known variables (independent variables). The result is obtained by using the known variables and processed according to a defined equation. The linear regression can be simple or multiple. In the case of a multiple regression, the value of the variable searched is dependent on more than one independent variable. It is the oldest methodology and has the advantage of being quickly "learned" and more easily interpreted.
- *TREE-BASED MODEL* is based on a sequence of operations that involve simple YES or NO or IF-THEN-ELSE decisions. After exhausting all logical questioning, we arrive at the result of the established sequence, which is very direct and easy to interpret. The advantage is that it can process non-linear situations.
- *NEURAL NETWORKS* mirror the functioning of the human brain, in which neurons communicate and exchange messages with each other. The concept was adapted for Artificial Intelligence under the name of *Artificial Neural Networks (ANN)*, as shown below.
- *ANN (Artificial Neural Networks)* or simply *NN (Neural Networks)* is nothing more than a computer system that simulates the functioning of the human brain. It is based on a set of nodes or units connected to each other, called artificial neurons, which simulate the neurons of the human brain. Each connection, like a synapse in the human brain, can transmit signals to other neurons. Each artificial neuron receives a signal, processes, and passes the result to others. The resulting signal from the processing is a real number, computed by means of some non-linear function, which represents the processing of the several inputs. The connections are called edges. The neurons and the connections (edges) are assigned weights, which are adjusted during the "learning" process. The weights can increase or decrease the value of the signal in the connections. Neurons may be assigned thresholds. Depending on these thresholds, a neuron may or may not transmit the received signal. Typically, neurons are organized in layers. Different layers can process different changes in the input signals. Input signals propagate

from the first layer (input layer) to the last layer (output layer), probably after traversing the layers multiple times. Figure 14.2 schematizes these concepts.

For a Neural Network to work, it must initially go through a learning or "training" period. This training consists of feeding the algorithm or model with input data, knowing the results previously. In this procedure, associations are formed with probabilistically found weights. These values are stored in the data structure of the network itself. The training process, starting with a certain known example, is usually conducted by determining the differences between the result found after processing (the prediction found), and the known value, which is the target to be reached. This is the error. The net then adjusts its weights associated with connections from defined learning rules, using the error as an adjustment parameter in succession, until the neural net can produce a response that approaches the expected value for each interaction. Based on certain criteria, training is terminated after several rounds. This process is known as Supervised Learning.

Artificial neurons communicate with each other according to several patterns, allowing the output of one neuron to serve as input for another. The network that is formed ends up being a large branch in which the values of the connections have weights, as already mentioned. These weights indicate the degree of influence of a neuron over the others. The result of the processing is provided by the output neurons that compose the neural network. To arrive at the final result, it is necessary to weigh the sum of all the weights of input neurons, weighted based on the weight of each of the connections of these neurons. A bias value is added to the result of this sum. This weighted sum is referred to as activation. The weighted sum is then subjected to an activation function, usually a nonlinear function, to produce the result. The initial data is usually external data, such as an image or documents. The expected result is the recognition of an image or document submitted to the network.

The process that comprises computing the sum of all input signals of a neuron according to the signals received from the outputs of the neurons that preceded it, considering the relative weight of each connection, is called *Propagation Function*. This result may still be subject to the addition of a value indicative of the weighting bias.

Neurons are organized into multiple layers, especially in *Deep Learning*, which is the case for structures composed of multiple layers. Between two layers, there

Fig. 14.2 Example of an ANN FeedForward Neural Network

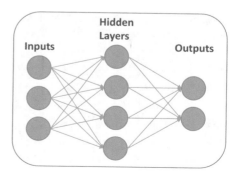

may be multiple possible connections. A node may be connected to all other nodes in its layer and all nodes in the next layer. Nodes can be pooled, so that a group of nodes in one layer is connected to a single node in another layer, thus reducing the number of connections in a layer. Networks of neurons with a single direction of connection are so-called feedforward networks (Fig. 14.2), i.e., networks that advance in one direction only. There are, however, networks that allow connections between neurons on the same or previous layers. These are the recurrent networks, as shown in Fig. 14.3.

The parameters used in AI or ML are of two types: the Hyperparameters, parameters defined as immutable, that is, defined a priori by the programmer before the "training" of the model. The other parameters are defined during the system *training* process.

Learning is the procedure used by the model to adapt the parameters so that the results are as satisfactory as possible. This is achieved by observation and involves changing the weights and eventually the thresholds of the network to improve the results of the model adopted. This process ends only when the possible interactions can no longer reduce the error rate in a useful way for the system. Normally, the error will never be zero. After the learning process, if the error remains high, the model should be redesigned. In practice, this is done by defining the *Cost Function*, a factor that is evaluated throughout the learning process. As long as the error is decreasing, "learning" continues.

A *Cost Function* is a technique used to evaluate the performance of an algorithm or model. It uses predictable input and output values and analyzes what the error value is after the values have been processed by the chosen algorithm or function. The "cost" is often defined by a statistic, whose value can only be estimated.

The learning process can be understood as a progressive optimization procedure that, associated with statistical procedures, is used to estimate results. There are three possible procedures used in the learning phase:

- *Supervised Learning*, already described above, in which sets of known input and output parameters are used. The learning process consists of obtaining results in

Fig. 14.3 Example of an ANN Feedback Neural Network

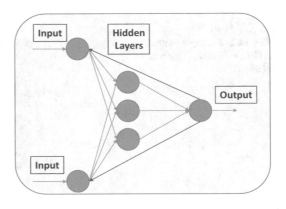

the output that are consistent with the input values. In this case, the Cost Function is used to correct incorrect outputs.

- *Unsupervised Learning*, which applies in situations where there are no known sets of data inputs and outputs to promote "learning." These are cases where you want to look for an unknown pattern within a large data set. The process, in this case, consists in clustering a set of elements according to some determined pattern, trying to characterize it according to some known statistical distribution.
- *Reinforcement Learning* is a procedure that aims to adjust the network through actions in order to reduce the cumulative cost of learning. At each moment the agent provokes an action, the environment generates an observation and the "instantaneous cost" resulting from this action, according to well-established rules. At this point, the agent decides whether to explore new actions that justify the cost, or to experiment returning to previous learning to accelerate the process. It is as if trying an innovation to see if it is worthwhile, a complex mathematical process according to a decision-making process developed by Markov (MDP – Markov Decision Process).
- *Self-Learning* is a system with only one input (condition/situation) and one output (action/behavior). It is not yet a fully autonomous system. In situations where the program must identify objects or images, it is fed with a large amount of data in order to create a neural network applicable to the situation. These systems need a large processing capacity for working with a very large volume of data, besides consuming a long processing time. To develop image recognition, for example, the system must be fed with a large volume of images of the object to be identified and a large volume of images that have nothing to do with the object in question. This process may require days or weeks to be "trained" before it can be used. But the process does not stop there and must continue to be refined until it reaches a level considered ideal. The literature compares this process to learning a language. One can spend years learning the grammar, verb conjugation, and syntax of a language or simply live for a while in a country that adopts the language one wants to learn. Certainly, the second alternative is much faster than the first.

For all that has been said and learned, we can realize that both Artificial Intelligence and its by-product, Machine Learning, are logical-arithmetic processes that emulate our way of thinking and making decisions based on a large volume of information that is stored. Over time, they allow us to act spontaneously, almost "without thinking," and make decisions that are processed by our brain "almost" automatically or intuitively. The processes, models, and statistical distributions are basic tools for the development of this science based entirely on mathematical knowledge. The machine seeks to classify data and fit them into some previously known model (stored) and process this data according to the logical-arithmetic processing that constitutes its essence. What is the limit to all this? Only time and the capacity of the processors can give us an answer with absolute probabilistic certainty.

Here is worth a note for those who have completed reading the previous chapters. In them, we learned that hardware is literally shrinking, as well as becoming more standardized and integrated, while the software is advancing in every way, taking full control of all the functions necessary for the perfect functioning of a digital cellular communication network. This shows, when reading about AI and ML, how much these concepts have been fully incorporated into 5G. This progression only tends to grow and allow 5G to become more "intelligent" and autonomous, depending less and less on human interference for its operation.

Recommended Reading

Title	Complement
Evolution of cellular technology	
Evolution of the Telecommunications Industry Into Internet Age	Matin Fransman University of Edinburgh
Evolution of Cellular Technology	Pearson 0137033117 (https://cdn.ttgtmedia.com/searchTelecom/downloads/SearchTelecom_Fundamentals_of_LTE_Chapter_1.pdf)
Mobile Broadband Explosion – The 3GPP Wireless Evolution	Rysavy Research for 4G Americas – August 2013
LTE to 5G: Cellular and Broadband Innovation	Rysavy Research for 5G Americas – August 2017
LTE – Long Term Evolution	
LTE The UMTS Long Term Evolution – From Theory to Practice	Stefania Sesia, Issam Toufik and Matthew Baker – Second Edition Wiley
LTE in a Nutshell – The Physical Layer	White Paper telesystem Innovation (2010)
LTE Overview	www.tutorialspoint.com/lte/lte_overview.htm
RF Planning and Optimization for LTE Networks	Mohammad S. Sharawi – CRC PressEditors: L. Song and J. Shen
An Introduction to LTE: LTE, LTE-Advanced, SAE, VoLTE and 4G Mobile Communications, 2nd Edition	Christopher Cox
LTE Technology Introduction	Rohde & Schwartz
Understanding LTE and its Performance – Network Architecture and Protocols Chapter 2	T. Ali-Yahiya, Springer Science+Business Media, LLC 2011
What is RRH (Remote Radio Head), How it is connected to BBU (Base Band Unit)?	www.techplayon.com/rrh-remote-radio-head-connected-bbu-base-band-unit/
Software-defined Open Architecture for Front- and Backhaul in 5G Mobile Networks	volker Jungnickel, Kai Habel, Michael Parker, Stuart Walker, Carlos Bock, Jordi Ferrer Riera, Victor Marques, David Levi

(continued)

© The Author(s), under exclusive license to Springer Nature Switzerland AG 2023
J. L. Frauendorf, É. Almeida de Souza, *The Architectural and Technological Revolution of 5G*, https://doi.org/10.1007/978-3-031-10650-7

Table 1 (continued)

Title	Complement
SAE Architecture	https://en.wikipedia.org/wiki/System_Architecture_Evolution
4G and 5G Migration and Interworking Strategy	White Paper – Cisco External – Version No.0.2 August 19,2020 – Rajaneesh Shetty, Prakash Suthar, Vivek Agarwal, Anil Jangam
5G Service-Based Architecture (SBA)	EventHelix Oct 20, 2018 (https://medium.com/5g-nr/5g-service-based-architecture-sba-47900b0ded0a)
High-level Architecture from '2G' to '5G'	@3g4gUK
Evolution of the UTRAN Architecture	Markus Bauer, Peter Schefczik, Michael Soellner, Wilfried Speltacker – Lucent Technologies, Germany
The LTE Network Architecture – A comprehensive tutorial	Stratigic White Paper – Alcatel Lucent
LTE for UMTS: OFDMA and SC-FDMA Based Radio Access -Chapter 6 LTE Radio Protocol	Antii Toskala and Woonhee Hwang – Edited by Harri Holma and Antti Toskala – John Wiley & Sons, Ltd.
Chapter 2 – LTE Physical layer	LTE Air Interface Training Manual – Huawei Technologies Co. Ltd.
Protocol Signaling Procedures in LTE	RADISYS White Paper – by V. Srinivasa Rao and Rambabu Gajula
LTE: Transmission and Reception Concepts	TELCO Intelligence in Telecommunications
Modulation	
Understanding Modern Digital Modulation Techniques	Electronic Design 2012-01-23 – by Louis E. Frenzel
Digital Modulation schemes – OQPSK – Offset Quadrature Phase-shift keying	Shahid Beheshti Univ. of Tehran Department of Electrical Engineering – October 2014
Multiple-Access Technology of Choice in 3GPP LTE	Ibikunle Frank, Dike Ike, Ajayi Jimi, Onasoga Kayode – Indonesian Journal of Electrical Engineering and Informatics (IJEEI) Vol.1 No. 3 September 2013
LTE and the Evolution to 4G Wireless: Design and Measurements Challenges – Chapter 2 Air Interface Concepts	Moray Rumney by Agilent Technologies – John Wiley & Sons
Digital Modulation	David Tipper – Associate Professor – Department of Information Science and Telecommunication University of Pittsburg
Practical Guide to Radio-Frequency Analysis and Design – Chapter 4 Radio Frequency Modulation – Digital Phase Modulation: BPSK, QPSK, DQPSK	https://www.allaboutcircuits.com/textbook/radio-frequency-analysis-design/radio-frequency-modulation/digital-phase-modulation-bpsk-qpsk-dqpsk/
5G	
5G for dummies	Kalyan Sundhar and Lawrence C. Miller – Ixia Special Edition
Understanding 5G – A Practical Guide to Deploying and Operating 5G Networks	VIAVI Solutions First Edition 2019

(continued)

Table 1 (continued)

Title	Complement
Understanding 5G – A Practical Guide to Deploying and Operating 5G Networks	VIAVI Solutions second Edition 2021
5G Mobile Networks: A Systems Approach	Larry Peterson and Oguz sunay
Transition Toward Open & Interoperable Networks	5G Americas White Paper – November 2020
Capacity and Cost for 5G Networks in Dense Urban Areas	Dave Wisely – University of Surrey – April 2019
Estimating the mid-band spectrum needs in the 2025–2030 time frame – Global Outlook	GSMA / Report by Coleago Consulting Ltd. – July 2021
Connected World – An evolution in connectivity beyond the 5G revolution	McKinsey Global Institute – Discussion Paper – February 2020
Flying to the Clouds: The Evolution of the 5G Radio Access Network – Chapter 3	Glauco E. Gonçalves, Guto L. Santos, Leylane Ferreira, Élisson da S. Rocha, Lubnnia MF de Souza, André LC Moreira, Judith Kelner and Djamel sadok
5G NR Logical Architecture and its Functional Splits	Parallel Wireless White Paper
5G Standalone architecture	Samsung – Technical White Paper – January 2021
Multi-Vendor 5G Core Networks: The Case for a Disaggregated Control Plane	Gabriel Brown, Principal Analyst, Heavy Readings White Paper / Oracle Communications – July 2021
Network Slicing and 5G Future Shock	Jennifer P. Clark, Principal Analyst, Heavy Readings White Paper / Ciena / Blueplanet – January 2020
5G & Beyond for Dummies	Larry Miller and Jessy Cavazos – Keysight Technologies – 2022
Massive MIMO for New Radio	Samsung Technical White Paper – December 2020
Non-orthogonal Multiple Access and Massive MIMO for Improved Spectrum Efficiency	Anritsu White Paper – Massaki Fuse and Ken Shioiri
Open RAN: The Long Journey from Supporting Act to Lead Role	Kester Mann, Director, Consumer and Connectivity, CCS Insight White Paper prepared for interdigital – November 2021
5G and Shannon's Law	fate Khanifar and Sarvesh Mati – WAVEFORM – White Paper – May 2021
IP and WiFi networks	
Network Handout	Bruno Michel Pera – UNIVAP – version 3/2021
IP Networks – Architecture, Protocols and Applications	Evandro Luís Brandão Gomes – INATEL
The Wi-Fi Evolution	Jaidev Sharma – White Paper QORVO – March 2020
A Guide to Wi-Fi 6E – Wi-Fi 6 in the 6 GHz Band	White Paper LITEPOINT, A Teradyne Company – 2021
IP and WiFi networks	
Artificial Intelligence – intelligent systems	tutorialspoint White Paper – 2015

(continued)

Table 1 (continued)

Title	Complement
Machine Learning Basics: An Illustrated Guide for Non-Technical Readers	guidebook dataiku
3 Enterprise Trends Driving AI Into Everyday Use: 2022 and Beyond	EBOOK – Dataiku
A Case Study on Protocol Stack Integration for 3GPP LTE Evolved Node B	Fabbryccio A.C.M. Cardoso, Felipe A.P. Figueiredo, Rafael Vilela and João Paulo Miranda—Center for Research and Development in Telecommunications – CPqD, Campinas S.P. Brazil.

Index

Printed in the United States
by Baker & Taylor Publisher Services